WORKS IN PROGRESS

W O R K S
IN PROGRESS

ALVIN ROSENBAUM

DIANE MADDEX AND
GRETCHEN SMITH MUI
PHOTO EDITORS

FOREWORD BY ERIC DELONY

POMEGRANATE ARTBOOKS
SAN FRANCISCO

LIBRARY OF CONGRESS
CATALOGING-IN-PUBLICATION DATA

Rosenbaum, Alvin.
Works in progress / Alvin Rosenbaum
   p.   cm.
ISBN 0-87654-069-8
1. Structural engineering.   2. Buildings — Design and
construction.   I. Title.
TA633.R66   1994   94-7912
624 — dc20   CIP

PUBLISHED BY POMEGRANATE ARTBOOKS

Box 6099, Rohnert Park, California 94927–6099

PRODUCED BY ARCHETYPE PRESS, INC.,
WASHINGTON, D.C.

Project Director: Diane Maddex
Editor: Gretchen Smith Mui
Designer: Robert Wiser

PHOTOGRAPHS

Endpapers: Construction of Shasta Dam, Shasta County,
California, 1942.
Pages 3, 4, and 5: Construction of the Hirshhorn Museum
and Sculpture Garden (1970–74, Skidmore, Owings and
Merrill) on the Mall in Washington, D.C.
Pages 6–7: Erection of defense housing in Erie,
Pennsylvania, in 1941.
Page 9: Hand of the Statue of Liberty being enlarged in
plaster during work at Gaget, Gauthier, and Company, Paris,
1876–81.

Printed in Singapore

10   9   8   7   6   5   4   3

CONTENTS

Foreword   12

Preface   15

Introduction   16

THE MECHANICAL AGE

U.S. Treasury Building   26

Washington Monument   28

Central Park   30

U.S. Capitol Dome   32

Prospect Park   34

Brooklyn Bridge   36

Chicago Fire   38

Philadelphia City Hall   40

Statue of Liberty   42

Pension Building   44

Sagrada Familia   46

Library of Congress   48

Eiffel Tower   50

Washington Square Arch   52

Biltmore Estate   54

Trans-Siberian Railroad   56

Cathedral of St. John the Divine   58

THE CITY BEAUTIFUL

World's Columbian Exposition   62

General Noble Redwood Tree House   64

Barn Raising   68

Boston Subway   70

New York Public Library   72

Flatiron Building   74

Grand Central Terminal   76

Pennsylvania Station   78

Times Tower   80

Coney Island   82

San Francisco Earthquake   84

Panama Canal   86

Titanic   88

Crooked River Bridge   90

Lincoln Memorial   92

Soddie   94

Set of *Intolerance*   96

Electric Trolley Cars   98

Key West Highway   100

Paramount Pictures Gate   102

Route 66   104

George Washington Bridge   106

Mount Rushmore   108

Louisville Palace Theater   110

BUILDING BIG

Goodyear Airdock   114

Chrysler Building   116

Empire State Building   118

Radio City Music Hall   120

U.S. Botanic Garden   122

Hoover Dam   124

Arthurdale   126

Golden Gate Bridge   128

Grand Coulee Dam   132

Greenbelt   134

Rockefeller Center Ice Rink   136

Ringling Brothers and Barnum and Bailey Circus   138

Fallingwater   140

Johnson Wax Administration Building   142

Sharecropper's Cabin   144

Trylon and Perisphere   146

Set of *Mr. Smith Goes to Washington*   148

Jefferson Memorial   150

Adobe Restoration   152

Nazi Party Headquarters   154

Manzanar   156

THE POSTWAR ERA

Levittown   160

White House Renovation   162

Fontainebleau Hotel   164

Disneyland   166

Sydney Opera House   168

TWA Terminal   170

Dulles International Airport   172

U.S. Air Force Academy Chapel   174

Gateway Arch   176

Marina City   178

Astrodome   180

John Hancock Center   182

U.S. Pavilion, Expo 67   184

Sears Tower   186

Vietnam Veterans Memorial   188

Big Ben Restoration   190

Pyramide du Louvre   192

Channel Tunnel   194

Oriole Park at Camden Yards   196

U.S. Holocaust Memorial Museum   198

Endeavour   200

Statue of Freedom Restoration   202

Illustration Credits   204

Index   205

FOREWORD

When photography was invented, the United States was just 65 years old. The camera arrived early in our process of taming nature and subjecting the wilderness and its inhabitants to the national doctrine of manifest destiny. The person who made the first photograph in this country was the man who invented the telegraph, and its popularity spread just as quickly. While in France promoting his invention in 1839, Samuel F. B. Morse met Louis Daguerre and brought back with him a daguerreotype, the "mirror with a memory."

By the mid-1850s the process of developing unlimited paper prints from a single negative had been perfected. The wet collodion negative of the 1880s was followed by the dry collodion negative, which was placed in direct contact with a light-sensitive paper and produced a same-size paper print called a contact. Glass-plate negatives provided an even more crystal-clear view of the world—an objective report of works in progress.

A photograph, however, is not only a direct impression of an object. It is also an image focused by the lens of the larger world outside the camera. Photographs of structures under construction have been made since the invention of photography, but systematic documentation of the construction process did not occur until the costs and convenience allowed a professional photographer to be included as part of the team. In the first half of the nineteenth century the client or engineer hired an artist to illustrate the construction. The results were magnificent port-folios of lithographs or engravings now coveted by museums and collectors. When printing technology made possible the direct transfer of screened halftone photographs to the printed page, illustrated portfolios were supplanted by photographic portfolios originally intended for news or engineering periodicals. Enhanced with a series of finely executed drawings and a description of the structures, they were made into lavish bound portfolios for presentation to project backers or used as advertising for engineering services.

As head and principal architect of the Historic American Engineering Record, a National Park Service program charged with creating an archive of drawings, photographs, and histories of America's engineering and industrial achievements, I have had the enviable opportunity for the last 25 years of working with photographs old and new. For the past 25 years HAER has been racing to capture the vestiges of nineteenth- and twentieth-century American engineering and public works before they are lost. The role of our documentary photography is to save the visual evidence and record what is significant, both common and ordinary structures as well as particularly fine and well-designed examples. HAER's contemporary photographs contribute to the overall message of a site, providing the visual part of the analysis and illustrating a logical sequence that, along with measured drawings and written histories, allows

13

us to form a logical conclusion about how something works. Historic photographs, where we are fortunate to find them, give depth to the site and provide an opportunity to go back in time and compare what was seen then with what survives today.

The value of photographs as documents of construction was brought home in the 1980s, when the National Park Service restored the aging Statue of Liberty in New York Harbor. A gift from France to commemorate America's birth, the figure that greets immigrants and travelers at the gateway to America was intended to be complete for the country's centennial in 1876. However, because of a lack of subscriptions to pay for America's part of the bargain—the stone base—it was not completed for another 10 years.

A century later the "Mother of Exiles" was in need of repair. When a French-American team assembled to plan the statue's restoration, surprisingly few photographs of its construction were found. Because the 1880s was the apogee of photography's popularity, one would have thought that a systematic series of photographs would have been taken of this monumental icon. Mountains of research for the restoration were compiled on both sides of the Atlantic, but the hoped-for mother lode of historic construction photographs, essential to renovation planning, never materialized.

To ensure that this shortcoming would not be repeated a hundred years later, when another generation might have to plan another restoration, the National Park Service requested that HAER methodically record the work. Important steps in the construction process were identified: the statue before restoration, constructing the scaffolding, removing the torch, strengthening the raised right arm, gutting the base, removing the wrought-iron straps supporting the skin and replacing them with stainless steel, reattaching the replicated torch, disassembling the scaffolding, and restoring the statue and grounds. In addition to the thousands of photographs taken by project contractors, more than 250 archival-quality construction photographs, both color and black-and-white, were taken by HAER for permanent deposit in the photographic collections of the Library of Congress.

During the last 150 years that construction photography has been available to us, attitudes toward it have shifted several times: one generation manipulates its negatives and prints to create artistic impressions, while the next rejects this approach in favor of objective, utilitarian reports. In architectural photography today the public usually sees only the finished structure—forever losing the opportunity to understand exactly what went into a building's construction. *Works in Progress* allows us the chance, at least with the structures included here, to step back into time and into the process by which both landmarks and common buildings enter our world.

Those of us involved in historic preservation use photography for yet another purpose. When people ask what the value of documentation is, I answer that we are in the amenity business, the responsibility of saving the very best handed down to us so that we can pass it on for the future. Saving structures of fine materials, humanly scaled proportions, notable craftsmanship, and varied textures enhances the quality of life and maintains familiar surroundings. In places where historic architectural and cultural resources are lacking, attitudes supporting good community design may also be absent. Such values are especially needed in America, where we tend to throw away the past, build the expedient, pursue the quick profit, and, in the process, desecrate the countryside. In this age of instant gratification, dramatic photographs such as the views here provide a link with the past as well as deeper insight into and appreciation for the soaring human imagination and activity portrayed.

Eric DeLony, Chief
Historic American Engineering Record

PREFACE

I have been fascinated with construction since boyhood. I grew up in a Usonian house designed by Frank Lloyd Wright for my parents in Alabama in 1940. The vocabulary of our house was obvious—a kit-of-parts that logically combined all the elements of its structure. Early on I began to look at other structures inside out.

In New York one day in the winter of 1981 my friend and colleague Nina Castelli Sundell and I imagined a book called *Works in Progress*, a visual chronicle of landmark buildings under construction. We realized then that we would need to find photographs that captured buildings at the stage where their form was recognizable but not so far along that their structural elements were completely covered. Our musings over lunch went no further, each of us moving on to other projects.

A dozen years later I remembered the idea and discussed it with Diane Maddex, president of Archetype Press. She helped me develop the concept further and took it to Thomas Burke of Pomegranate Artbooks, who offered the resources for me to begin.

In the abstract the idea of construction photography seemed both straightforward and not difficult to research. In reality I discovered that construction photographs of the 85 structures we selected were scattered among dozens of archives and in some cases did not exist at all. Ford Peatross, curator of architecture, design, and engineering collections, Library of Congress, suggested some likely sources, and we began the long and fascinating detective work necessary to assemble the photographs included here. Diane Maddex and Gretchen Smith Mui, editor at Archetype Press, performed a large measure of this work, for which I am grateful.

Robert L. Wiser and Marc Alain Meadows designed the book to be as grand and heroic as the structures themselves. Robert M. Vogel, former curator of engineering and industry, National Museum of American History, Smithsonian Institution, and Eric DeLony, chief, Historic American Engineering Record, both read the manuscript and provided invaluable suggestions and corrections. Any errors that remain are entirely my own.

Alvin Rosenbaum

Every day sidewalk superintendents pass along construction fences to witness a seemingly random but actually artful and carefully planned choreography of workers, materials, and equipment—a factory of effort rising from nowhere. At first work appears to progress at a snail's pace as earth is scooped from the ground into an endless line of vehicles, to be carted away to fill someone else's hole. Piercing the ground to what seems an impossible depth, pile drivers create an insane racket as steel is hammered into the earth to help form piers that ultimately will support a building's foundation. Months pass as this work proceeds, the entire project apparently only a catacomb that will never rise to the street's level.

In time the first floors begin to emerge, a floor added each few days, the whole reaching the sky in only several weeks. Toward the end the pace slows again as the building is finished and outfitted with fixtures, as the lobby's marble is set and the elevators are installed. All of it seems standard and routine—a thousand buildings requiring the same standard skills of a thousand workers. But building begins on paper and not in the ground, where calculation and the laws of physics direct the process.

Abandoning the tried-and-true, architects and engineers continuously attempt to find a better way to get the job done. With every breakthrough in the building sciences a problem has to be overcome: first, in principle, working with a slide rule or a computer; second, by the

design of a prototype structure and the manufacture of new, adapted, or custom materials; and then on the construction site, as all the imagination, courage, wisdom, calculation, and precision workmanship are finally brought together in the creation of a building, dam, bridge, vessel, tower, or monument.

For the invention of a civilization, construction is the means that assists transport, provides family shelter, and aids commerce. Construction also facilitates recreation and entertainment, governance and learning, and glorifies, memorializes, and helps make important what is both useful and beautiful in society.

THE INDUSTRIAL AGE

Over the past century construction has been industrialized, requiring that a wide range of materials and skills be brought together from disparate disciplines and sources. The sequence of discovery and the perfection of processes have provided the necessary prerequisites for innovation. Certainly the burning of coal in the smelting of iron was an essential development that preceded the Bessemer process (1865), which industrialized the conversion of iron into steel and provided a practical manufacturing method to supply production quantities of the material for large-scale construction projects. Before the Bessemer process, only cast and wrought iron—the former lacking tensile strength and ductility and the latter lacking the overall

SCULPTOR GUTZON BORGLUM SET OUT IN 1927 TO ETCH IN STONE FOUR PRESIDENTS HE THOUGHT HAD THEMSELVES MOST SHAPED AMERICA. HIS HEROIC MOUNT RUSHMORE WAS ALMOST COMPLETED WHEN HE DIED IN 1941.

strength of steel—were available for structural purposes, and only a few places could produce pieces large enough for monumental structures. The most important of these early sites was the blast furnace at Coalbrookdale in west-central England. Ironmaster Abraham Darby (1678–1717) planned a 100-foot span to cross the Severn River with five parallel semicircular arches, cross-braced for strength to support a 24-foot-wide roadway 55 feet above the river. Each of the ribs of the bridge was cast in 70-foot pieces weighing 38 tons, floated down the river on barges, hoisted into place with block and tackle, and joined together with cast-iron bolts. Iron Bridge was completed in 1779 and stands today as a monument to the industrial revolution.

One hundred years after Iron Bridge was erected, Chicago architect William LeBaron Jenney

AT THE BEGINNING OF THE TWENTIETH CENTURY MANY RURAL AREAS STILL RELIED ON HORSEPOWER FOR HEAVY CONSTRUCTION PROJECTS. WORKERS BEGIN CONSTRUCTION OF A STREET IN IOLA, KANSAS, ABOUT 1900.

(1832–1907) designed what many regard as the first skyscraper: the Home Insurance Company Building (1885), a milestone made possible by the development of skeletal framing, the availability of cheap steel, and the invention of the elevator. All this became feasible through commercial necessity: the rising cost of land in downtown Chicago.

Between Iron Bridge and the Home Insurance Building—icons of late eighteenth-century and late nineteenth-century industrialized construction—came the development of the steam engine and the railroad, the evolution of manufactured building materials, the standardization of interchangeable parts, the use of balloon framing in housing, and, by the end of the nineteenth century, the introduction of electricity as a power source in factories and of gasoline-driven engines on construction sites.

Industrialization of construction had a remarkable effect on both design and engineering. With the first use of cast iron in bridges, structural elements were revealed. Cast iron replaced wood and stone, which in turn were later replaced by steel as the prevalent structural material in construction. Now stone is cut like salami into thin slices and applied as a skin to the steel structure, often creating the appearance of solidity beneath the surface where only a skeleton exists.

As construction turned to steel, structural frameworks were increasingly hidden. An early and leading proponent of the idea of edifice as artifice was Daniel H. Burnham (1846–1912), whose firm, Burnham and Root, designed the 21-story Masonic Temple (1890) in Chicago, then the tallest building in the world. Burnham soon became chief of construction for the World's Columbian Exposition of 1893, whose buildings captured the essence of the new construction ethos: buildings framed in lumber or steel were clad in a material called "staff," a mixture of cement, plaster, and hemp that was troweled onto wooden lath, molded, and sculpted to give the appearance of lasting through the ages and yet concocted to survive only for the fair's six-month run.

In creating these false impressions, Burnham followed the lead of Prince Grigori Potemkin. Attempting to persuade Catherine the Great to annex the Crimea in 1787, Potemkin erected fake facades of buildings to look like villages along Catherine's travel route, so that she would believe that the area was filled with new subjects for her rule.

By the beginning of the twentieth century Burnham and Charles McKim (1847–1909) of New York paved the way for architects across the country to recreate the "White City" of the 1893 exposition, constructing monuments to civic pride and grand ambitions—enduring tributes to democratic ideals made permanent, neoclassical buildings that appeared to have been created by the ancients yet somehow transported into the modern era.

As a practical matter, buildings such as the New York Public Library (1897–1911) and Union Station (1903–8) in Washington, D.C., simply could not have been created as they were without structural steel. The irony of Burnham's White City was that by using the most modern of construction methods, historic styles that once were executed with the toil of millions of hours of slave labor and hundreds of skilled artisans could be built in a matter of months, using essentially unskilled labor. Ornamental elements could be produced in sections on a lath using a pattern or cast in terra cotta; window moldings, door frames, niches, statuary, gargoyles, Corinthian columns, moldings, pediments, and other devices could be created on shop floors in long or short runs and shipped to the construction site by rail or steamer.

During the period 1880 to 1920, marked by massive migrations and displacements of populations and the fall of monarchies around the world, the architectural trend was a belief—an invented tradition—that the Old World's hallowed halls could somehow be cloned and remade into pieces of new cultures, offering stability in places gone awry by the progress of industrial technology, bigness, dislocation, and unprecedented long-term prosperity and economic growth in Europe and North America.

Parallel to the rise of the faux in American building construction was the introduction of advertising and consumerism in the American economy, both products of industrialized methods grafted to the introduction of modern salesmanship. Just as the Lincoln Memorial (1915–22) and other classical structures in Washington (proposed by the McMillan Commission, whose members included Burnham and McKim) created an impression of solidity, seriousness, and ageless enduring values, the sculpting, packaging, and advertising of Larkin Soap served precisely the same purpose. The principal proposition of both architecture and advertising was a kind of universal optimism, a mix of biblical virtue and a belief in the future. Over time this proposition formed the basis of an emerging motion picture industry, its sets for spectacles becoming parodies of Burnham's White City—grandeur created in plaster to look real, recalling Potemkin's villages of flimsy scaffolding holding thin facades, all front and no back.

While American architects were first experimenting with structural steel, Europeans were pioneering the use of reinforced concrete, a structural material more frequently given to honest expression. Pioneered by Joseph Monier in Paris in the 1860s for basins and tubs, concrete reinforced with iron was later extended for use experimentally as sleepers for railroad tracks and walkways and eventually used for bridges, arches, pipes, and even floors and roofs. In 1900 Swiss engineer Robert Maillart (1872–1940) first used reinforced concrete in load-bearing structural elements for slab floors in factories. Architects soon began using concrete sculpturally, creating free-form buildings such as the Einstein Tower (1919–21) in Potsdam,

designed by Erich Mendelsohn (1887–1953), and eventually the rhapsodic TWA Terminal (1956–62) at Idlewild Airport (now John F. Kennedy International Airport) in New York, designed by Eero Saarinen (1910–61).

REVOLUTION IN LABOR

Over the recent history of construction, changes in the application of labor have paralleled the evolution in methods and materials. At the beginning of the nineteenth century only a few architect-engineers in America were available to plan and supervise public works. Benjamin H. Latrobe (1764–1820), credited as America's first professional architect, designed Philadelphia's first waterworks in 1800 and later important parts of both the U.S. Capitol and the White House. Not until the building boom after the Civil War was specialization introduced; architecture and engineering began to be treated as separate disciplines and engineering specialties were created. Most house construction during America's colonial era was by farmer-mechanics who carefully crafted intricate notches to create mortise-and-tenon joints from hewn timbers and fastened these with dowels and wooden pegs. Each structure was individualized and hand-crafted from mostly native materials.

The introduction of balloon framing—the use of lightweight, saw-milled lumber for plates and studs, joists, rafters, headers, and other components held together by manufactured nails—transformed residential construction. With it and the advance of urbanization came the transformation of a labor force, as highly skilled self-employed jacks-of-all-trades went to work in city factories and gave up country life. As factory workers were engaged in performing repetitive tasks, construction work maintained an important link with the past, its tasks requiring greater degrees of skill, flexibility, and versatility.

JOHN A. ROEBLING, WHO PERFECTED THE TECHNOLOGY OF STEEL CABLES, MADE POSSIBLE MODERN SUSPENSION BRIDGES SUCH AS THE GOLDEN GATE BRIDGE (1933–37), WHICH SPANS THE SAN FRANCISCO BAY.

EERO SAARINEN'S DESIGN FOR THE KRESGE AUDITORIUM (1955) AT THE MASSACHUSETTS INSTITUTE OF TECHNOLOGY
WAS A PROLOGUE TO HIS DESIGN FOR THE TWA TERMINAL AT IDLEWILD AIRPORT THE FOLLOWING YEAR.

While building has evolved to require specialized skills—in carpenters, crane operators, masons, plumbers, electricians, and others—these crafts have overlapped, necessitating a broad knowledge and the ability to work as part of a team on interrelated tasks. This collaboration in the building trades is joined by a tradition of independence, as employment shifts from job to job across the lives of construction workers.

Inevitably manufactured systems have replaced skilled craft in many areas of construction, requiring less fashioning of parts and more factory-like assembly work on the job. Design, prototyping, and the creation of a kit-of-parts have become the sine qua non of productivity, in which customization of a standard configuration has replaced truly custom work in most areas of architecture and construction. Increasingly, many architects today are left with only

the task of conceiving a design for the surface, the entry, and the lobby of a large building. Virtually everything else is left to the engineer, the builder, and real estate and business facility specialists to optimize the revenue and utility of the space contained within.

CENTURY OF PROGRESS

Advances in construction methods followed advances in technology in its golden age, a century of progress beginning in 1850. Society stood in awe of the first grand engineering feats, from London's Crystal Palace (1851) and the canal at Suez (1871), to the Allegheny Portage Railroad (1834) and the Brooklyn Bridge (1869–83), designed by John A. Roebling (1806–69) and his son Washington A. Roebling (1837–1926). Society applauded and went on to

invest in roads, bridges and tunnels, methods of transport, and impressive monuments to its own successes. For the dedication of the Suez Canal, Giuseppe Verdi was commissioned by the British to write an opera, creating *Aida* in the process. The results of an engineer's genius were cause for celebration: the Eiffel Tower (1889), designed by Alexandre-Gustave Eiffel (1832–1923), was seen by 32 million visitors during the Universal Exposition's run in Paris in 1889. In the United States lavish temples of technology were constructed for the centennial fair in Philadelphia in 1876, in Chicago in 1893, in St. Louis in 1904, in San Francisco in 1915, in Chicago in 1933, and in New York in 1939.

As futuristic buildings advanced progress at world's fairs, authors and critics raised engineering on a pedestal. Henry Adams stood before a great dynamo and declared its power greater than that of religion or the work of the Renaissance painters; Rudyard Kipling glorified bridge builders, while H. G. Wells and Jules Verne spun utopian tales graced by mechanical genius. As technology began to create problems in overcrowded cities and routinized factory work, Aldous Huxley's *Brave New World* (1932) and later Charles Chaplin in *Modern Times* (1936) portrayed technology and the engineer as evil — a force of totalitarianism, not freedom.

At the end of the golden era, when innovation was almost completely given over to military pursuits, engineering's prestige began to wither. Large public works — dams and power plants, superhighways, and other megastructures — were fraught with opponents seeking to protect the environment and slow a progress perceived to create inequities in society. For those works in progress caught in the crossfire, such as nuclear power plants, billions of dollars were wasted as these hulks of concrete and rebar stood on the landscape as ugly half-built reminders that engineering never advances too far ahead of the society it serves.

After more than a century of good press, construction as a bellwether of society's good health began to be cast as a harbinger of society's problems. The speculation of the 1980s that used federally guaranteed investment to produce an endless parade of postmodern containers for high-tech expansion finally came to a close. Routine building of large civic and commercial structures may now be history as banks, insurance companies, and government agencies have finally restrained the pace of growth in the name of prudence, either to protect the environment or because they are broke.

Advances in technology over more than a century were made evident by huge construction sites; the process of building was often a public spectacle that lasted months or years, employing masses of workers, promising to beautify the landscape and magnify our pride of place. Now technological advances are hidden on the space shuttle and within silicon chips, their presentation through fiber-optic cables sprayed onto pixilated screens in our homes and mobile offices. What had been riveted is now joined by code; foundations are now poured by strategic planners onto spreadsheets.

The construction of large public works and buildings has sometimes been an act of great civil obedience, like landmark legislation — rays of vision shining through an uninspired bureaucracy. Photographs of works in progress remember the effort that was made to create these landmarks large and small — tributes to optimism and to bigness — all the while glorifying history. Progressive construction photography has often been part of the construction process of large buildings and bridges since the 1860s, providing a legal record for liability claims and other disputes, technical documentation to assist with any potential problems with the building, and a historical account of the construction itself. The focus now and in the years to come will be on the preservation of such structures as important links to our past and as monuments to both the idea and the craft of building as the greatest expression of a civilization.

THE MECHA

NICAL AGE

1836 – 71

WASHINGTON, D.C.

The Treasury Building was constructed piecemeal over a 35-year period under the supervision of five successive architects. It was begun in 1836 by Robert Mills (1781 – 1855), whose competition design was selected by President Andrew Jackson. Between 1836 and 1842 Mills supervised construction of the spine and central wing, whose major architectural feature is the dramatic 466-foot-long Ionic colonnade along Fifteenth Street; the interior boasts one of Mills's distinctive fireproof double flying staircases. An 1855 design by Thomas U. Walter (1804 – 87) called for imposing projecting porticoes on the south, west, and north facades. This design was the basis for completing the exterior of the building, although later architects made changes inside. Ammi B. Young (1798 – 1874), named supervising architect of the Treasury Department in 1852, oversaw construction of the south and west wings. Monolithic granite columns for the west wing exterior were cut from the quarry on the diagonal; the photograph, dated September 16, 1861, was taken during this period. When Young retired in 1862, Isaiah Rogers (1800 – 69) took over the west facade; he served as supervising architect for two years, 1863 to 1865, and designed and patented the cast-iron walls for four burglar-proof vaults in the south wing. The north wing was designed by Alfred B. Mullett (1834 – 90), who followed the Greek Revival exterior pattern established by his predecessors but designed his interiors in the Renaissance Revival style.

27

1845–54, 1876–84

WASHINGTON, D.C.

In 1833 the Washington National Monument Society was founded to erect a monument whose dimensions and magnificence would symbolize America's gratitude to its first president. The society intended to raise the projected $1 million by public subscriptions of $1 per American, but four years later only $25,000 had been raised. A design competition was held in 1836, when the site had not yet been selected, but none of the entries was deemed appropriate. Robert Mills's design, finally chosen in 1845, consisted of a 600-foot Egyptian obelisk on a colonnaded base. In 1848 the society asked Mills and James Renwick (1818–95) to modify the design and limit the height to 500 feet. The cornerstone was laid in June 1848. By 1851 the shaft rose 80 feet. As construction progressed, each row of stonework became more difficult to lift into place. Work on the monument came to a halt in March 1854, when it had reached about 170 feet. A stone donated by the Vatican was stolen by members of the American Party, an anti-immigration, anti-Catholic group known as the Know-Nothings. The obelisk remained a truncated stump until 1876, when public interest in the monument was rekindled by the centennial of the American Revolution. After Congress assumed funding for the project, the Army Corps of Engineers reinforced the base and eventually completed the monument, now capped with a 50-foot-tall pyramid designed by Lt. Col. Thomas Lincoln Casey. Dedicated by President Chester A. Arthur on February 20, 1885, the monument to George Washington remains, at 555 feet, the world's tallest masonry structure.

1857 – 73

NEW YORK CITY

In 1844 *New York Evening Post* editor William Cullen Bryant called for a large, open pleasure ground for New Yorkers. With the help of landscape architect Andrew Jackson Downing, a cheap, mostly vacant, relatively flat, swampy, pestilential 840-acre plot was surveyed and purchased in 1856. The following year Bryant and Washington Irving helped organize a competition for the design of the first municipal park in America. Landscape architect Frederick Law Olmsted (1822 – 1903) and architect Calvert Vaux (1824 – 95) won from among the 33 entrants with a plan called Greensward. For Olmsted and Vaux the barren tract was an empty canvas, an opportunity to mold from nothingness a garden of magnificent proportions, recreating vistas more common to the English countryside than to urban America. The naturalness of Central Park is purely an illusion crafted from Olmsted's imagination, a blending of prospects and places of refuge with its woods and open spaces, tangles and glades, promontories, craggy heights, ponds and lakes. This 1863 glass-plate photograph shows the terrace at Bethesda Fountain (facing south), designed by Emma Stebbins. The national influence of Olmsted, known today as the father of American landscape architecture, is seen clearly in the grand estates of the Gilded Age, the grounds of the U.S. Capitol, and in hundreds of college campuses and municipal parks across the United States. Over the years several large structures have intruded on the park, many contrary to Olmsted's vision. In 1980 the Central Park Conservancy was founded to preserve the park in the Olmsted tradition but adapt it to modern needs.

1860–63

WASHINGTON, D.C.

Because of a pressing need for more space, the U.S. Capitol (1793) was expanded beginning in 1850 from plans drawn up by Thomas U. Walter, architect of the Capitol. The new wings he proposed created a problem: the original wooden dome, designed by Charles Bulfinch (1763–1844), was out of proportion with the rest of the newly grandiose building. Walter thus designed a new dome, based on St. Peter's in Rome, made of cast iron. As the Civil War began, work on the Capitol dome and the House and Senate wings continued by order of President Lincoln: "When the people see the dome rising," he said, "it will be a sign that we intend the union to go on." In 1860–61 the foundry Janes, Fowler, Kirtland and Company of Brooklyn delivered to the site 1.3 million pounds of cast-iron parts to be bolted together for a dome supported by 36 continuous trussed ribs. The total weight of the ironwork was nearly 4,500 tons. The dome—287 feet high, 135 feet in diameter at the base, and 88 feet at the cupola, later topped with Thomas Crawford's 19½-foot bronze statue *Freedom*— is the only one of this size to be added to an existing building. During the Civil War the Union Army's Aeronautic Corps appropriated the Mall for wartime purposes. The "balloon corps"— observers in tethered balloons who conducted aerial reconnaissance of Confederate encampments across the Potomac—set up its hydrogen gas generators there, a view captured in this photograph by Mathew Brady (1823–96).

33

1865–71

BROOKLYN, NEW YORK

Because Prospect Park, unlike Central Park, cannot be traversed by automobile and because it is in Brooklyn rather than Manhattan, it remains more remote and more pristine. The two parks share the same designers—Olmsted and Vaux. But in this case their services were retained without a competition, thus allowing significantly more freedom for a design to evolve from the site itself. The first assemblage of land for the park was bisected by Flatbush Avenue and was rejected by the architects. As a result of land swaps after the Civil War, the "prospect" of Prospect Park, along with a large reservoir, was eliminated. The eventual site of 350 acres was created from land acquired in 1859 mostly from Edward Litchfield. Lt. Gen. Egbert L. Viele led the survey work for the park beginning in the summer of 1865. As its name suggests, Prospect Park was conceived as a place of sylvan vistas and solitary walks as much as a gathering place for Brooklynites. Olmsted and Vaux's concept for the park followed a three-part scheme of meadow, woods, and water. Prospect Park was considered the more successful design of the two New York parks designed by Olmsted and Vaux because the designers were given more time and freedom to develop a plan for spectacular vistas and a number of exquisite architectural amenities, including classical works—bridges, statuary, fountains, terraces, lampposts, and park benches—by such artists and architects as John H. Duncan, Frederick MacMonnies, Thomas Eakins, William O. Donovan, and McKim, Mead and White.

34

BROOKLYN BRIDGE

1869–83

NEW YORK CITY–
BROOKLYN, NEW YORK

The Brooklyn Bridge was the longest single-span suspension bridge in the world when built. Designed in the 1860s by John A. Roebling (1806–69), engineer of the Cincinnati Suspension Bridge (1867), it began as a private enterprise that immediately was marked by corruption and tragedy: Roebling died as a result of an accident at the bridge site, and his son Washington A. Roebling (1837–1926) took over. While working in one of the caissons, he contracted caisson disease; he was left partially paralyzed, unable to leave his room but able to view the bridge's progress from his bedroom window. The two Gothic-arch towers sit on foundations formed of wooden caissons sunk 45 feet deep on the Brooklyn side and 75 feet deep on the New York side. The Roeblings perfected the engineering and construction principles that made long-span suspension bridges practical. Their work on the Brooklyn Bridge revolutionized suspension bridge construction throughout the world. A significant contribution was the development of cable-making machinery that spun steel wire into cables on site. The four main cables from which the roadway is suspended are 16 inches in diameter, are composed of 5,282 iron wires each, and are 3,578 feet long. On the main span 208 suspenders hold up the roadway, which is 85 feet wide and 1,595 feet long between the towers. After the bridge opened, wagons, carriages, and horse cars at both ends created unparalleled vehicular chaos. For a while, the only reasonable way to get across the bridge was by foot or on the bridge's cable cars.

CHICAGO FIRE

 1871

CHICAGO

In 1833 George Washington Snow (1797–1870) introduced balloon-frame construction to Chicago, a method using vertical studs that replaced the more expensive post-and-beam construction of the time. Innovation in construction, including mass production and the use of machine-made parts (such as factory-produced nails), became a hallmark of Chicago because of its rapid growth: in 1850 the city's population was only 29,963, but by 1870 it had jumped to 300,000. Downtown Chicago was a hodgepodge of shanties lining muddy lanes among a jumble of lumberyards, warehouses, offices, factories, boarding houses, and saloons. A fire on October 8–9, 1871, destroyed the heart of Chicago, leaving 90,000 people homeless and resulting in more than $200 million in property damage. The disaster stimulated major changes in the city plan, permitting developers to plan wider streets and more substantial structures and organize city blocks in a way that increased the city's capacity for growth. The street level was raised, resulting in an underground infra-structure, parts of which are still used. As the process of rebuilding Chicago began, the corner of Randolph and Market streets was documented in this photograph. The fire coincided with the industrialization of construction, creating condi-tions ripe for innovation—the erection of tall, efficient buildings and the use of inexpensive construction methods—and leading to the birth of modern architecture.

1871–1901

PHILADELPHIA

Philadelphia's city hall, delayed by political rivalries and massive corruption, required 30 years to build—a monument to civic pride out of control. The seven-story Second Empire–style structure is the largest municipal government building in the United States, perhaps the world. The Center Square site, selected by William Penn in 1683 as one of the city's original five public places, was the site of Benjamin H. Latrobe's Greek Revival waterworks pumping station. Two design competitions were held for the building—one in 1860 and one in 1868—and John McArthur, Jr. (1823–90), a Philadelphia architect, won both. Thomas U. Walter served as consultant and is credited with much of the building's detailing. Actual construction began in 1871, and the building was occupied in stages after 1877. During construction the state-appointed city hall building commission was ejected from the building by order of the mayor and police and reinstated only by the Pennsylvania Supreme Court. Both McArthur and Walter died before the building was completed; it was finished under the supervision of W. Bleddyn Powell. The first story is granite; the upper stories are brick faced with white marble, while the highest stages of the tower are iron. The city hall is capped by a 548-foot tower with a statue of William Penn, lifted into place in 1894 and for years the highest allowed construction in the city. The statue, as well as the building's extensive bas reliefs and caryatids, was sculpted by Alexander Milne Calder (1846–1923), a Scottish immigrant whose grandson and namesake became one of the most notable American sculptors.

STATUE OF LIBERTY

1874 – 86

NEW YORK HARBOR

In 1874, soon after French sculptor Frédéric-Auguste Bartholdi (1834–1904) began to create his work *Liberty Enlightening the World,* he retained Alexandre-Gustave Eiffel (1832–1923) to design the statue's wrought-iron skeleton. Bartholdi began with a four-foot clay model. Working with skilled craftsmen at Gaget, Gauthier, and Company in Paris, he made three progressive plaster enlargements, each involving nearly 10,000 measurements. Wooden frames were then built to follow the exact contours of the plaster model. Thin copper sheets—one one-hundredth of an inch thick—next were hammered into shape over wooden frames. These were held together by iron straps and fastened to the skeleton. When the colossal work was completed, its 350 lightweight pieces were shipped across the Atlantic Ocean on the French frigate *Isère.* Liberty finally rose on Bedloe's Island in New York Harbor, on a pedestal designed by Richard Morris Hunt (1827–95), in 1886. From foundation to torch she is 305 feet; from base to torch, 151 feet; from head to toe, 111 feet. Her index finger is 8 feet long, her mouth is 3 feet wide, her head is 30 feet high, and her right arm is 42 feet long. The original flame that beckoned the huddled masses to America's shore was replaced in 1916 by a lantern of amber glass resculpted by Gutzon Borglum (1867–1941). The statue was restored for the American bicentennial, a task whose engineering complexity almost rivaled that of the original construction.

PENSION BUILDING

1882-87

WASHINGTON, D.C.

Although it used 15.5 million red bricks instead of travertine and stucco, Gen. Montgomery C. Meigs's design for the Pension Building was inspired by the Palazzo Farnese in Rome, the largest and most famous of all Renaissance palaces. Notable in the construction were eight Corinthian columns 75 feet tall and 8 feet in diameter that were later plastered and painted to imitate Siena marble, a 159-foot-tall interior court with an iron-trussed roof, and a 1,200-foot terra-cotta frieze by Caspar Buberl (1834–99) depicting Union Army experiences during the Civil War. The mammoth columns are positioned at each end of the inner court and flanked by a rising series of arcades. Meigs (1816–92), a former quartermaster general who supervised the construction of many Washington structures, introduced several innovations in the building: double-pane glass reduced summer heat; most corridors were eliminated to provide a usable work area of 80 percent (versus 50 or 60 percent then common); and a central clerestory admitted ample natural light and good ventilation. The center court of the Pension Building, which can accommodate thousands of people at one time, has been used for more than a dozen inaugural balls, including President William McKinley's second inaugural ball in 1901 (seen in the bottom photograph). During the building's tenure as the Pension Bureau, 1885–1926, more than $8 billion was paid out to 2.7 million veterans and their families for service in wars from the American Revolution to the Civil War. The building is now home to the National Building Museum.

SAGRADA FAMILIA

1884–

BARCELONA, SPAIN

Some artists simply do not fit any of the neat categories devised by historians to explain a place and time. The work of Antonio Gaudí y Cornet (1852–1926) may have paralleled Art Nouveau yet it was quite different: an individualistic, impassioned, and often compulsive rhapsody of organic imagery with spirals of flora and fauna. His Casa Mila (1905) in Barcelona, known as the "quarry," looks as though it were made of clay. His Sagrada Familia (Expiatory Church of the Holy Family), modified over the years to reflect his developing style, goes beyond the influence of Art Nouveau and actually foreshadows the Expressionist movement in twentieth-century architecture. The crypt (1883–87) is strictly neo-Gothic, the apse walls and finials (1891–93) are a rather free Gothic, and the first transept facade (Nativity) (1893–1903) is in the evolving modernist style. The second transept (Passion) was constructed after 1954 according to his 1917 drawings. Structurally, Gaudí's studies for the nave (1898–1925) were most important and reveal his rationale for keeping the masonry structure self-supporting within. The four striking towers were completed only after his death. Work has continued sporadically over the years, but construction will not be finished for another generation.

LIBRARY OF CONGRESS

1886–97

WASHINGTON, D.C.

Following the vision of Thomas Jefferson, who provided the first library for Congress, Ainsworth Rand Spofford created today's Library of Congress. It was Spofford who first proposed that the library become the repository for books and periodicals under the copyright law. He served as librarian from 1864 to 1897, when the library opened at its present site. During his tenure the collection's holdings grew from 80,000 to more than 1 million items. In the 1873 design competition for the building, 27 architects submitted designs. Proposals ranged from a Victorian Gothic structure that looked like the Houses of Parliament to a massive Romanesque pile resembling New York's first Grand Central Depot. During the competition, won by the architects John L. Smithmeyer (1832–1908) and Paul J. Pelz (1841–1918), who proposed a monumental Beaux Arts design, debate about the site continued; a group from Congress insisted that the library be placed next to the old Botanic Garden near Judiciary Square to be closer to the White House. Construction began in 1886 and was completed 11 years later. This construction photograph of the famous dome over the Main Reading Room was taken on October 18, 1894, by Levin C. Handy.

OCT 18. 1894

EIFFEL TOWER

PARIS

A decade after Alexandre-Gustave Eiffel devised the wrought-iron skeleton on which the Statue of Liberty is hung, he designed a tower for the 1889 Universal Exposition in Paris—perhaps the only world-renowned landmark named for its designer. The event was organized to celebrate the centennial of the French Revolution and to display French leadership in industry, the arts, and commerce. Among the artifacts of the exposition is the 984-foot Eiffel Tower, which was condemned by Paris artists and intellectuals as a structure with the appearance of a machine but which broke new ground in engineering. As a result of the tower's geometry, the structure is extremely light; the pressure on the ground is no more than that of a person sitting on a chair. The perforated arches between the tower's four legs serve no support function but are simply decorative, a concession by Eiffel to reassure the public that the whole structure would not collapse. The tower was built in only a few months at a modest cost.

WASHINGTON SQUARE ARCH

1890-92

NEW YORK CITY

The original Centennial Arch over lower Fifth
Avenue, a temporary structure designed by
Stanford White (1853–1906) of McKim, Mead
and White and paid for by the residents of the
Washington Square neighborhood, was
constructed of plaster and wood in 1889 to commemorate the centennial of George Washington's
first inauguration. Painted white and decorated
with stucco wreaths, it was topped with an
eight-foot statue of Washington. Because of the
arch's enormous popularity, a committee was
formed to raise $150,000 for construction of a
permanent marble structure in Washington
Square, slightly south of the original location.
The monument was financed by subscriptions to
a benefit concert by pianist Jan Paderewski and
by gifts from William Rhinelander Stewart,
shown in a top hat in this 1892 photograph. The
Roman arch's great success helped establish the
classical ideal for virtually all the commemorative
architecture in New York during the era. The
inscription reads: "Erected by the People of the
City of New York." The statue of Washington on
the west pier of the arch was sculpted by
Alexander Stirling Calder (1870–1945), father
of Alexander Calder.

BILTMORE ESTATE

1890-95

ASHEVILLE, NORTH CAROLINA

Biltmore, the largest of America's large houses, was designed by Richard Morris Hunt for George Washington Vanderbilt, a grandson of "Commodore" Cornelius Vanderbilt. Hunt, known for his palatial private homes for wealthy clients, including The Breakers and Marble House in Newport, Rhode Island, drew inspiration for this 255-room mansion from early sixteenth-century French châteaux in the Loire Valley such as Blois, Chenonceaux, and Chambord. The building's architectural tour de force is a double-spiraled towered stair, a copy of a version at Blois that spirals in the opposite direction. A self-sufficient estate of 125,000 acres, supplying its own food and income and much of its power, Biltmore encompassed formal gardens, a farm, a dairy (now a winery), a village, and a forest. A railroad spur three miles long was built to bring in tons of Indiana limestone for the walls. The house, the repository of a rich collection of art and furnishings, also incorporated a complex heating system, elevators and dumbwaiters, an elaborate fire-alarm system, mechanical refrigeration, and plumbing. A portion of the estate—a 250-acre pleasure park and a series of gardens around the house—was landscaped by Frederick Law Olmsted; the vast wooded domain was managed experimentally by Gifford Pinchot (1865–1946), who later established the U.S. Forest Service. In 1898 the Biltmore Forest School was founded; it trained the first generation of American foresters in conservation techniques still applicable today. The estate, opened to the public in 1930, today encompasses 8,000 acres.

1891–98

RUSSIA, NORTH OF THE CAUCASUS MOUNTAINS

During the waning years of imperial Russia, public works ranged from massive construction projects in St. Petersburg to the 4,600-mile Trans-Siberian Railroad, which extended from Chelyabinsk in the Ural Mountains to Vladivostok on the Pacific coast. The railroad was the largest public works project in history, requiring more than 80,000 laborers, and the major economic and defense initiative in czarist Russia. The railroad included a dozen smaller regional efforts, including lines linking the provinces of Russian Caucasia—Armenia, Azerbaijan, Georgia, Daghestan, and Adzharistan—and terminating at Nakhichevan, near the Turkish and Persian borders. Into this obscure corner of the planet traveled the World's Transportation Commission, formed as the result of the World's Columbian Exposition of 1893, one of whose members was the photographer William Henry Jackson (1843–1942). The trip was not only a glamorous junket but also provided a unique setting for Jackson's travel photographs in an age when the Grand Tour was de rigueur for the upper classes. Jackson made this photograph of the construction of the line between Tzchenif and Zkhali in 1895 on the final leg of the 17-month trek. Although the railroad was crucial for troop transport during the Russo-Japanese War and World War I, the trains that followed this route were slow and had limited cargo capacity; only six trains per day—three in each direction—could be used. The railroad, the only means in and out of Siberia, was also intended to aid colonization, which proved an equally slow process.

56

CATHEDRAL OF ST. JOHN THE DIVINE

1892 –

NEW YORK CITY

The vision of an Episcopal church structure to rival the great cathedrals of England was introduced by the bishop of New York, John Henry Hobart, in 1828, but construction did not begin for another 64 years, on December 27, 1892, when the cornerstone was laid at 110th Street in Morningside Heights on the Upper West Side of Manhattan. The cathedral's design resulted from an 1889 competition that attracted nearly a hundred entries. The award went to Heins and La Farge, whose eclectic design was a composite of Romanesque, Byzantine, and Gothic forms. During excavation of the site, substantial remedial work was required when loose rock, underground springs, and compressible earth were encountered. The photograph shows a column being transported to the site in 1904. With the death of architect George L. Heins (1860–1907) and his replacement by Ralph Adams Cram (1863–1942), whose design was rejected in the original competition, Romanesque plans gave way to always-popular Gothic motifs. The work of Cram and Ferguson that began in 1911 has continued through successive generations of architects in the firm to this day, although the nave was completed before Cram's death. The cathedral is 603 feet long with twin towers 266½ feet high, enclosing a floor area of 121,000 square feet. The sanctuary seats 6,000 people and can accommodate an additional 2,000 standing.

THE CITY

BEAUTIFUL

WORLD'S COLUMBIAN EXPOSITION

1893

CHICAGO

For Daniel H. Burnham (1846–1912) the 1871 fire that ravaged Chicago was a blessing: "There are [now] no buildings possessing either historical or picturesque value which must be sacrificed...." The city's vacant lots provided a blank canvas for him to apply his vision of a new Chicago. Two decades later Burnham, as chief of construction for the 1893 exposition celebrating the four hundredth anniversary of Columbus's arrival in America, was handed another opportunity to execute his ideas about city planning and neoclassical architecture. These gave rise to the concept of the City Beautiful, a noble place of classical proportions that celebrates the arts, industry, and progress and elevates the human spirit. Burnham's orchestration of the fair's buildings, landscaping, statuary, and ornament introduced the idea of ensemble architecture. The White City, comprising the Court of Honor buildings and the Palace of Fine Arts, was a product of Beaux Arts influences, reflecting Roman elements in the triumphal arches, arcades, domes, and sculpture. In this photograph, taken by Frances Benjamin Johnston (1864–1952), workers finish a scrollwork panel. Constructed for temporary use and finished with a plaster of paris compound that looked like white marble, most of the 150 fair buildings were destroyed when the gates closed. More than 27 million people visited the exposition during its six-month run. The fair's impact on civic design was profound: it influenced the McMillan Commission's design of the Washington Mall as well as the hundreds of neoclassical buildings constructed throughout the country.

GENERAL NOBLE REDWOOD TREE HOUSE

1893

CHICAGO AND WASHINGTON, D.C.

One of the most sensational exhibits of the 1893 World's Columbian Exposition was the General Noble Redwood Tree House, a 50-foot section of a 2,000-year-old giant sequoia. Erected as part of a U.S. government exhibit, the tree was cut in the former General Grant National Park (now Kings Canyon National Park), Three Rivers, California. A scaffold was built around the 300-foot-tall tree some 50 feet above the ground. Cutting through the 26-foot diameter required 20 cutters working for one week. The stump was then hollowed out, and the shell, including the bark, was cut into approximately 30 sections, to be reassembled, with an interior circular staircase, at the Chicago fair. In 1894 the tree was moved to the grounds of the old Agriculture Building on the Mall in Washington, D.C. At that time it was made weatherproof by the addition of a peaked roof with four dormer windows, all covered with redwood shingles. The tree house remained on the Mall until 1932, when it was placed in storage at the Agriculture Department's experimental farm in Arlington, Virginia. In 1940 the farm was transferred to the army for part of the Pentagon grounds, at which time the tree was probably burned. Ironically, the tree was named for John Willock Noble, who as secretary of the interior from 1889 to 1893 worked to ensure passage of an 1891 law to preserve millions of acres of western forests owned by the federal government.

1895

RAINY RIVER, MINNESOTA

Barn building on the frontier required a whole community to help with the framing, given the size and weight of the timbers and the extent of the task. Some barns measured 50 by 75 feet, although typically they were about half that size. Even so, a single framing timber could weigh hundreds of pounds. The timber was hewn for mortise-and-tenon connections, and the whole was then covered with sheathing; this 1895 photograph shows a late example of this type of construction. Lighter-weight balloon construction for barns became the norm in the Midwest after the Civil War, while New England farmers continued to build with massive timbers and central king posts designed to support gables. New England and Pennsylvania Dutch barns were built on hillsides (and thus called bank barns). The hay was stored on the upper ground floor, and cattle were herded into a basement below; the stock was accessible on the down-side of a slope. Another regional difference in barn construction emanated from the size of the farm. In New England, farms remained relatively small. In the wide-open spaces of the South and Midwest, farms grew as farming methods improved; Minnesota dairy farms were twice the size of those in Vermont. In addition, balloon construction encouraged large, flexible open spaces inside for equipment, animals, hay, and silage, whereas barns constructed of timbers were smaller and divided into even smaller spaces. Consequently, New England farms were made up of several small buildings in a quad-rangle, groupings that provided additional shelter from winter storms, while the large barns of the Midwest stood alone.

1895-97

BOSTON

Boston's Tremont Street Subway, America's first underground transit system, was built after years of debate among city officials, reformers, and trolley czars. Before the subway, it was said, congestion was so bad on Tremont Street at rush hour that pedestrians could walk from Scollay Square to Boylston Street on the tops of the stalled streetcars. Construction beneath Boston Common began on March 28, 1895, with plans adapted from the Budapest subway system. Excavation under Tremont Street at the corner of Boylston Street can be seen in this April 16, 1896, photograph. On September 1, 1897, the first subway trolley, crammed with 175 excited Bostonians, traveled the 1⅔-mile distance from Boylston Street to Park Street. The *Boston Globe* reported that more than 5,000 passengers made the trip each hour that day, "jammed like sardines and ... yelling like a jungle of wild animals." In its first year the line carried 50 million passengers. Reformer Sylvia Baxter raved about the subway's "white enameled walls, its brilliant electric illumination, its commodious stations ... everything as cleanly as the traditional Dutch housewives' kitchen," but *The Brickbuilder* complained that architecturally it was "about as enlivening and cheerful as a second-century catacomb." In 1914 the subway was extended to Kenmore Square, from which passengers took streetcars to the western suburbs. It also was renamed the Green Line, in honor of Frederick Law Olmsted's so-called Emerald Necklace, his park system connecting the city with outlying suburbs on the trolley line. The subway grew to include 85 stations and accommodate 206,000 passengers a day.

TREMONT ST AT COR OF BOYLSTON   Apr. 16 01

NEW YORK PUBLIC LIBRARY

1897 – 1911

NEW YORK CITY

"Whatever we now build … the law of historical development demands that it be Renaissance," noted Thomas Hastings at the turn of the century. Hastings (1860 – 1929) and his partner, John Merven Carrère (1858 – 1911), designed buildings in whatever style their clients desired so long as they were neoclassical and, preferably, Beaux Arts. In an 1897 competition their design for the New York Public Library was selected over that of McKim, Mead and White, the leading New York Classical Revival architects and the firm where both Carrère and Hastings had worked before opening their own practice in 1885. One of the last great nineteenth-century buildings, the library's facade was inspired by Claude Perrault's 1665 design for the colonnade of the east front of the Louvre. Remarkable for a New York building of this era, the $9 million library was set 75 feet back from Fifth Avenue, with plantings, rails, steps, seating alcoves, retreats, and sculpture providing a sympathetic setting. The New York Public Library was built at a time when Chicago architects were expressing the structure of office buildings built with steel framing. True to its own form, the library's architects used only white Vermont marble except in a few places, such as the ceiling moldings, where wood was used. The rear of the library, which faces Bryant Park, was designed with strong vertical elements that encased the library's heating system. The library's main portal is guarded by E. C. Potter's famous lions, which were rolled to their perches in 1911.

FLATIRON BUILDING

1901-02

NEW YORK CITY

The Fuller Building, more commonly known as the Flatiron Building, was the world's tallest habitable building—300 feet tall—at the time of its construction. Designed by Daniel H. Burnham for the Fuller Construction Company, it took its shape from the triangle formed by the intersection of Broadway and Fifth Avenue at Twenty-second Street. Although it is a steel-framed skyscraper, twice as high as any of its neighbors when built, Burnham clad it in limestone, subordinating function to aesthetics. The cost of construction was $2 million. Originally its design provoked shock and scorn, but today it is among New York's most beloved landmarks. To Brendan Gill it evokes not a flatiron but the Winged Victory of Samothrace, "poised with incomparable self-assurance on her windswept corner."

1903–13

NEW YORK CITY

During the Civil War the main railroad terminal
in New York was at Fourth Avenue and Twenty-
sixth Street, followed by the first Grand Central
Depot (1871). When the directors of the New York
Central and the New York, New Haven, and
Hartford railroads considered replacing the
station, they conceived of the project on a grand
scale. As the *New York Times* noted at the
dedication of the present structure, the terminal
was simply the keynote building of a massive
terminal city made possible by electric power:
"The scheme could not have been carried out—
it could not even have been conceived—in the
day of dirt and smoke and noise of the old steam
locomotive. The rock-bottom fact of the entire
enterprise is the electric motor, powerful, swift,
silent and clean." Tracks could now be safely
placed underground; open tracks, required for
ventilation for steam engines, were no longer
necessary. When the new Beaux Arts–style
Grand Central, designed by the firms of Reed and
Stem and Warren and Wetmore, was begun in
1903, what is now Park Avenue from Forty-fifth
to Fifty-sixth streets was one vast train yard. For
the new terminal two levels of tracks were sunk
below the street, and the surrounding 30-block
area became a real estate bonanza. Giant girders
spanned the sunken tracks, becoming the
foundations for a restored Park Avenue, cross-
town streets, and hundreds of new buildings.
The railroad's rights to the air space above
provided reclaimed "land" that was the most
valuable commercial property in the country.

PENNSYLVANIA STATION

1904-10

NEW YORK CITY

In Florence and Rome architects of the Renaissance built great buildings of stone — marble and granite — without the benefit of modern methods and materials. When structural steel was introduced into the engineering of large buildings, McKim, Mead, and White, the premier American architecture firm, incorporated this modern material with the classical forms of the Beaux Arts. One of the style's glories was Pennsylvania Station, a public building of giant piers, columns, and a coffered, vaulted ceiling of extraordinarily majestic proportions that rose 138 feet. The $112 million station's concourse, modeled after the Baths of Caracalla, was longer than the nave of St. Peter's. In 1963, despite public outcry, the station was demolished so that a new Madison Square Garden, relocated mostly below ground, could take its place. "The tragedy," wrote Ada Louise Huxtable in the *New York Times* that year, "is that our own times not only could not produce such a building but cannot even maintain it...." Through the old Pennsylvania Station, observed Vincent Scully, "one entered the city like a god. Perhaps it was really too much. One scuttles in now like a rat." The loss gave rise to New York City's landmarks ordinance and spawned thousands of other local laws that now protect historic structures throughout the United States.

TIMES TOWER

1905

NEW YORK CITY

Like the Flatiron Building, Times Tower, constructed for the *New York Times*, was sited on a triangular lot, for some the center of the universe: Longacre Square, transformed by this structure into Times Square. When completed, the building was New York's second tallest structure, surpassed only by the 32-story Park Row Building. Times Tower, designed by Cyrus Lazelle Warner Eidlitz (1853–1921), is known for two promotional devices derived from its unusual shape, both traditions that continue today. Beginning in 1908, on New Year's Eve a six-foot illuminated iron (later aluminum) Great Ball was lowered down the tower's 70-foot flagpole on the north face of the building, reaching its base at the stroke of midnight. Twenty years later the *Times* "Zipper," an electronic news sign consisting of 15,000 20-watt light bulbs controlled by a mechanical computer, announced Herbert Hoover's defeat of Al Smith, governor of New York, in the 1928 presidential election.

1905

BROOKLYN, NEW YORK

The early history of Coney Island as a seaside resort is marked by a series of devastating fires that destroyed each colony built on its shores. Coney Island House, the first hotel, was constructed in 1829 and later replaced by the Coney Island Pavilion and then Wyckoff's Hotel. In 1883 seven houses went up in smoke, and less than a decade later several blocks were turned to cinders, including Paul Bauer's Hotel. By the beginning of the twentieth century, Coney Island had developed into an enormous amusement park, lighted at night by a hundred thousand lights, a spectacle that rivaled or surpassed the expositions and world's fairs of the era. In fact, it was the great popularity of the Midway Plaisance, an amusement park at the 1893 World's Columbian Exposition, that spurred the development of Coney Island as an amusement center and influenced much of its architecture. Steeplechase Park opened in 1897, followed in 1903 by Luna Park, a fantastic melange of plaster-and-lath towers, and joined in 1905 by Dreamland, a project of William H. Reynolds, whose other ventures included the Chrysler Building. The photograph shows the Rocky Road to Dublin ride nearing completion. The park's timber buildings, strong winds, and inadequate fire-fighting equipment contributed to fires that plagued Coney Island for years. One night in May 1911 a fire destroyed its Dreamland midway, leaving only its world-famous Cyclone roller coaster. Luna Park and Steeplechase Park survived until the 1970s.

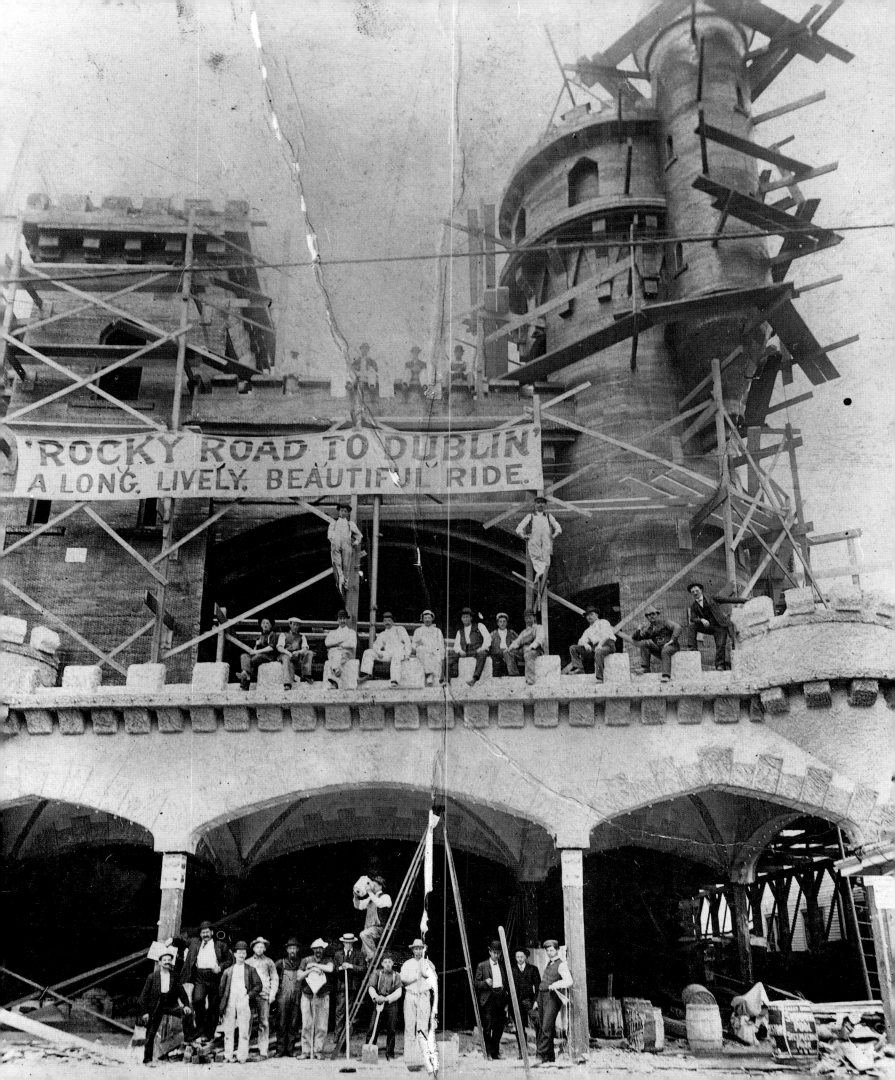

1906

SAN FRANCISCO

On April 18, 1906, the worst earthquake in American history destroyed 28,000 buildings in San Francisco's city center, an area of 4.7 square miles. Virtually everything of value was lost, including records in the county courthouse — birth, death, and marriage certificates and property and tax rolls—resulting in a generation of documentary confusion. One of the structures that could be stabilized after the quake was the Ferry Building, seen at left. Soldiers sent to guard the city engaged in looting, and distribution of government relief was predicated on the commanding general's order that "those who won't work shouldn't eat." Graft, unchecked among city officials before the earthquake, reached a peak during the rebuilding. One scandal resulted when most of the perfectly serviceable underground cable system used for downtown transportation was converted to an overhead trolley system; the ordinance was passed by the board of supervisors just after the earthquake — and on receipt of $200,000 in bribes. It permitted United Railroads to use the city's rebuilt streets, saving the company millions of dollars. Opportunities to rebuild San Francisco on the basis of new regional planning ideas were ignored. In 1905 Daniel H. Burnham of Chicago, the leading proponent of the City Beautiful movement, had developed an ambitious plan for San Francisco's growth to relieve congestion and provide a mix of parks, residences, and commercial development in a new regional scheme. The Burnham plan was a missed opportunity obscured by officials who publicly expressed no interest but privately were threatened by it.

1908–14

ISTHMUS OF PANAMA

In 1880 the French attempted and failed to build a sea-level canal across the Isthmus of Panama. Twenty-five years later President Theodore Roosevelt appointed a commission to consider the feasibility of building a canal at sea level versus one with locks. Eventually the commission recommended a sea-level canal. The plan had considerable support on Capitol Hill, but the problems with the sea-level approach were the same as those that had defeated the French: flooding and an unstable geology. Resisting the commission's findings, Roosevelt named a Second Isthmian Commission, which found in favor of the lift-lock proposal. After a long debate Congress approved the proposal. Work on the 51-mile canal was supervised by Col. G. W. Goethals, a U.S. Army engineer, and involved as many as 30,000 workers, who battled yellow fever and malaria as well as the heat. Labor problems, such as strikes by steam-shovel engineers, also added difficulty to the project. The canal cost about $380 million. The locks, designed by two civilian engineers, Edward Schildhaurer and Henry Goldmark, were built in pairs to accommodate two-way traffic: three pairs at Gatun, one pair at Pedro Miguel, and two pairs at Miraflores—twelve chambers in all. More than 5 million tons of cement were shipped from the United States to Panama for the concrete lock chambers. The Gatun locks, shown here, measured 110 feet wide, 1,000 feet long, and 90 feet deep and were divided by walls 60 feet thick. The excavation of the canal was a tremendous achievement for American engineering, comparable to digging a hole 16 feet square through the center of the earth.

87

TITANIC

1909–11

BELFAST

In 1907 Lord James Pirrie, a partner in the Harland and Wolff ship-building firm, and Bruce Ismay, managing director of the White Star Line, dined together to discuss an idea to ensure the line's domination in the international competition for transatlantic travelers — three huge luxury ocean liners. At its Belfast shipyards Harland and Wolff constructed giant gantry cranes and oversized docks specifically for these great ships. The second of these, the *Titanic,* was laid down in 1909. When launched on May 31, 1911, it was the largest ship afloat. Its eight decks rose the height of an 11-story building; it was 882 feet long and had a gross weight of 46,382 tons. Designed with a double bottom and 16 water-tight compartments extending the length of the keel, it was thought to be unsinkable. On April 10, 1912, the *Titanic* began its maiden voyage, sailing from Southampton for New York. On Sunday, April 14, at 11:40 p.m. it struck an ice-berg in the North Atlantic. The impact caused the steel plates along the six forward watertight compartments to buckle and the rivets to pop, allowing water to gush in. Because the compart-ments were not sealed at the top, the water flooded each in turn and poured into the rest of the ship. The *Titanic* sank two hours and 40 minutes later. Of the 2,223 passengers on board, 1,517 were drowned, including the ship's designer, Thomas Andrews. Among the survivors was Ismay, who climbed into the last lifeboat and was villified afterwards for his cowardice.

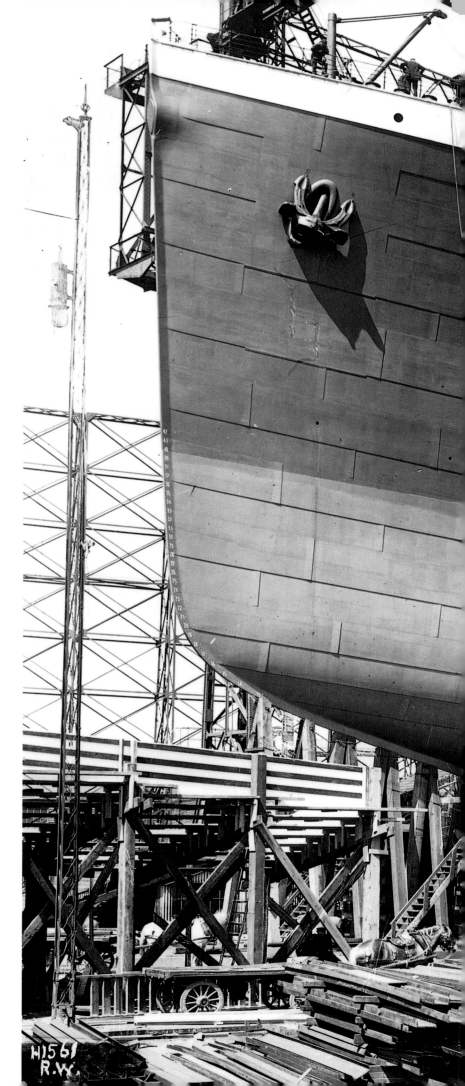

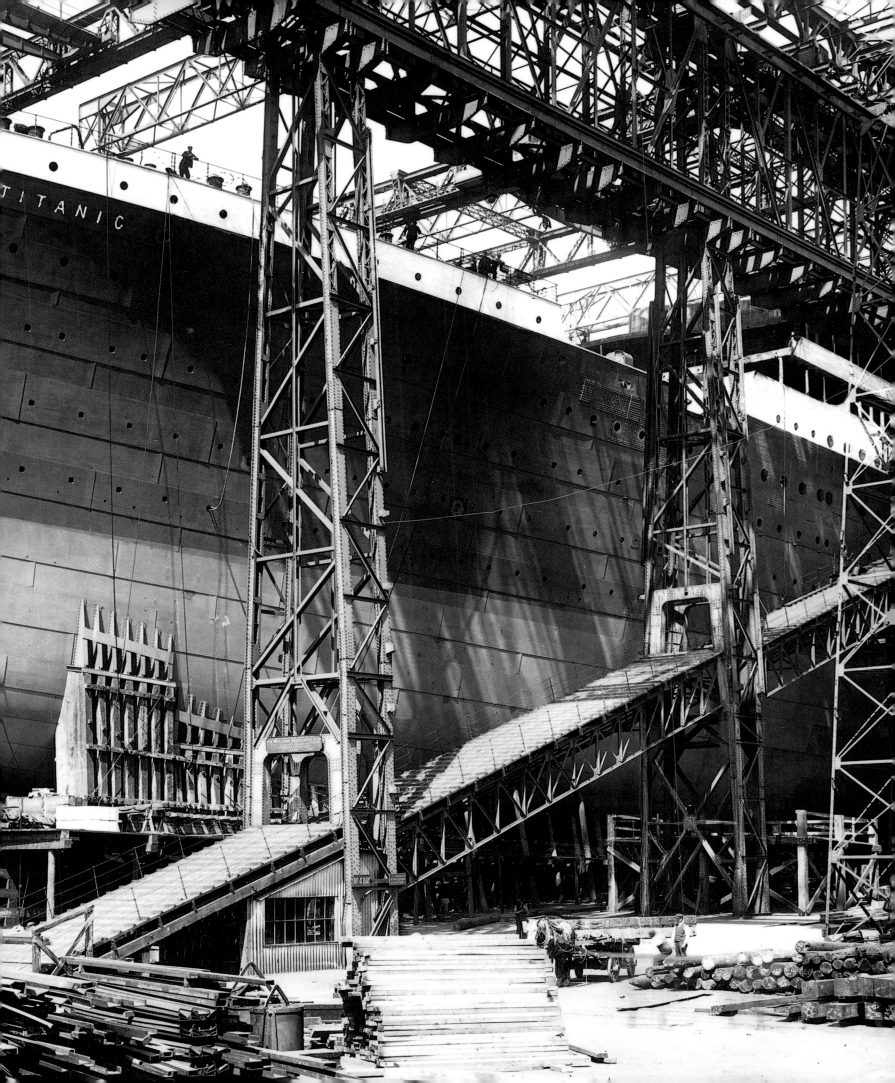

CROOKED RIVER BRIDGE

NEAR BEND, OREGON

The Oregon Trunk Railway in central Oregon
was built as a result of a fierce competitive battle
between James J. Hill of the Northern Pacific
Railroad and Edward H. Harriman of the Union
Pacific Railroad. The two empire builders'
decision to combine forces and build the railroad
as a joint venture in 1908 was a $90 million
gamble: the area had no industry and few
settlers. The Crooked River Bridge, part of the
Oregon Trunk, has a 69-foot-wide deck that
spans a gorge 340 feet wide at an elevation
320 feet above the river. When built it was the
highest arched structure in the United States and
one of the highest in the world. The photograph
shows workers "walking the plank" 348 feet
above the water as they proceed to place the
last segment of the span.

LINCOLN MEMORIAL

1915–22

WASHINGTON, D.C.

In 1902 Sen. James McMillan presented to the nation a plan to turn the nation's capital into "a civic work of art." His concept featured the development of the Mall to complete part of Pierre L'Enfant's vision, making it a vast recreation area with scenic vistas. Essential to the plan was land reclaimed from wetlands along the Potomac River and the construction of a memorial to President Lincoln. Republican Speaker of the House Joe Cannon thought it was a waste of money. "So long as I live I'll never let a memorial to Abraham Lincoln be erected in that God damned swamp," he declared. Despite resistance, building began in 1915 and continued until 1922. Carved from 28 blocks of Georgia marble, the seated figure of Lincoln was created by Daniel Chester French (1850–1931), renowned for his *Minute Man* and other public sculptures. The heroic left hand of Lincoln was lifted into place in January 1920. The classical temple designed by architect Henry Bacon (1866–1924), never envisioned by L'Enfant, stands at right angles to the Mall, counterbalancing the Capitol and the Washington Monument and anchoring Arlington Memorial Bridge, which leads south, to Robert E. Lee's home, Arlington. Some architectural critics have called the building's form inappropriate to its function as a memorial to the humble Lincoln, but the statue continues to draw and move those who visit it.

SODDIE

1916

NEBRASKA

Sod houses, called "soddies," were built by pioneers during settlement of the Great Plains after the Civil War and well into the twentieth century, as this 1916 photograph shows. Of the more than one million thrown up on sere land in areas such as Kansas and Nebraska, only a few are extant—most the victims of natural weathering and moves up to more permanent and stylish residences. Soddies were constructed of sod cut into blocks measuring approximately 1 by 3 by ½ feet, which were laid like bricks, grass side down, between forms, called stretchers. These foot-thick walls provided extraordinary insulation; to block drafts, they were plastered or covered with newspapers inside. The pitched roof was also laid with sod; tree trunks were used for the ridge pole and rafters. To build one sod house required 10 acres of sod. In the spring soddies were often covered with flowers, and cows were known to graze on the roofs. Whether a dugout or a near-mansion, said one settler, sod houses were "made without mortar, square, plumb, or greenbacks."

SET OF INTOLERANCE

1916

HOLLYWOOD, CALIFORNIA

During part of the year 1916 the tallest buildings on the Los Angeles skyline were the twin 165-foot towers of Babylon, designed by Walter L. Hall and constructed under the supervision of Frank Wortman for D. W. Griffith's film *Intolerance*. The set covered more than 100 acres and served as a stage for a scene requiring in excess of 4,000 extras. In *A History of Film Design* Léon Barsacq observed that *Intolerance* set a standard for American scenic design: "realistic sets and costumes for everything connected with the history and daily life of the United States, [but] imagination and freedom of inspiration [soared] as soon as the subject was a foreign country, or European or ancient history." Griffith made *Intolerance* after his box-office hit *Birth of a Nation* (1915), whose glorification of the Ku Klux Klan brought accusations of racial bigotry. In *Intolerance* Griffith sought to illuminate the universality of prejudice by depicting scenes from four periods in history. The spectacle of the "Fall of Babylon" depicted the betrayal of Prince Belshazzar to the Persians. The sets surpassed earlier film spectacles by Italian filmmakers who constructed sets that were enormous and extravagant but that failed to attract the public's attention. Because funds for their demolition were lacking, the massive Babylonian sets stood as a ruin at the corner of Sunset and Hollywood boulevards for many years.

ELECTRIC TROLLEY CARS

1921

PHILADELPHIA

During the transition from steam to electric power, public transportation was one of the first major industries to embrace the future. From the 1890s, when horse-drawn buses began disappearing, until the late 1920s electric trolley cars were the major form of urban transportation in the United States, connecting cities with their new suburban bedroom communities, known as streetcar suburbs. Local systems were often run by electric utilities, which also were interconnected throughout densely settled areas. At J. G. Brill Company in Philadelphia all the components for trolleys used in systems such as the local Frankford Elevated were manufactured and assembled: in one part of the factory, cane was woven for passenger seats; in another, curtains and shades for car windows were stitched. The factory produced thousands of trolleys, each finished in the central erecting shop. Eventually most cities found the electric trolley system inflexible to meet changing development patterns, so private passenger cars and rubber-tired city buses replaced trolleys in all but a few cities.

1923–28

KEY WEST, FLORIDA

The series of trestles, bridges, and viaducts that connects the Florida Keys to the mainland began in 1904 as a route built by Henry M. Flagler for the Florida East Coast Railroad, one of his holdings. Flagler invested $28 million in the project—156 miles of track—and hired thousands of workers for this huge undertaking. Conditions on the job site and at the work camps were execrable, and more than 700 laborers perished during construction of the railroad, which was completed in 1912. In 1923 Key West county commissioners adapted Flagler's idea and built a road for motor vehicles paralleling the railroad. This 170-mile "highway," which opened in 1928, relied on a roadway, bridges, and 40 miles of ferry service with toll charges. Construction of one of these bridges is shown in the photograph, dated May 27, 1926. After a hurricane damaged 41 miles of the track in 1935, the railroad decided not to repair or continue rail service. The state of Florida purchased the railroad right-of-way and bridge structures for $640,000 and built a new roadway on top of the railroad trestles, thus saving approximately $10 million in construction costs. The Overseas Highway (1936–38) was an extension of U.S. 1 and incorporated 44 bridges. Thirty-seven of them were replaced under the Keys Bridges Replacement Program (1973–83), whose innovations included precast segmental box girder construction, the first time this technique was applied to bridges as long as the new Seven Mile Bridge (35,716 feet).

101

CASTING OFFICE

SEPT 3 1926
9 AM

PARAMOUNT PICTURES GATE

1926

HOLLYWOOD, CALIFORNIA

An amalgam of several motion picture distribution companies—Famous Players-Lasky-Paramount—took over the sprawling production facility of United Studios at Van Ness and Melrose streets in 1925 and began an extensive remodeling program, including the construction of a new administration building, dressing rooms, and other buildings in a Spanish Mission style. When such contract players as Gary Cooper, Wallace Beery, Gloria Swanson, and Marlene Dietrich came to work on August 24, 1926, they were greeted by workers who had begun that day to build forms for an entrance to the studio. By September 10, according to daily photographic documentation, work on the classical arch had been completed—and an instant Hollywood icon had been created. In a short time the Paramount Gate came to represent the glamour, exclusiveness, and security surrounding the movies. In his 1932 remake of the movie *Extra Girl*, Cecil B. DeMille, the noted Paramount director, featured the Paramount Gate as a barricade against the horde of aspiring starlets who trooped to Hollywood from America's heartland for fame and fortune.

1926–38

CHICAGO TO LOS ANGELES

In *The Grapes of Wrath* (1939) John Steinbeck
called U.S. Highway 66, commonly known as
Route 66, "the mother road." The U.S. Highway
66 Association describes it as "a destination in
itself," 2,448 miles of highway from Chicago to
Los Angeles across eight states—Illinois, Missouri,
Oklahoma, Kansas, Texas, New Mexico, Arizona,
and California. Route 66 was a product of a
grass-roots movement for better roads and one
of the main arteries of the 1926 National Highway System. The head of the lobbying effort to
create the highway was Cyrus Stevens Avery, a
Tulsa oil man interested in getting goods across
the country quickly. Route 66 incorporated
sections of old trails at its eastern and western
ends but also created new routes, particularly in
Oklahoma. The road was paved first in California
and Illinois; in between, until the end of the
Great Depression, travelers had to struggle
through red clay, sand, gravel, high creeks, and
low bridges. In the 1930s, as construction progressed and became standardized, pavement
was laid in two lanes, each nine feet wide.
The concrete slabs were 6 to 10 inches thick,
depending on the subgrade, and reinforced with
steel bars. The highway was fully paved by 1938.
Winding its way through small towns across the
West, the road became a symbol of hope for
those who sought adventure and, during the
Depression, a better way of life. It also helped
spur development of the motel, the roadside
diner, and other highway trades. Northeast of
Oklahoma City, a photographer captured Route
66 being widened in the late 1940s.

105

GEORGE WASHINGTON BRIDGE

1927–31

NEW YORK–NEW JERSEY

In his top-floor apartment at the Carlyle Hotel in midtown Manhattan, Othmar H. Ammann (1879–1965) could move from room to room and catch a glimpse of many bridges—the George Washington, the Verrazano Narrows, the Bronx-Whitestone, the Throgs Neck, the Triborough, and the Bayonne—all products of his design and engineering finesse. He was proudest of the George Washington Bridge (originally called the Hudson River Bridge), his first big project, begun in 1927. The original facade of the suspension towers, as conceived by consulting architect Cass Gilbert (1859–1934), called for them to be clad in stone, a massive classical statement from the architect of the U.S. Supreme Court Building (1935). In the end the skeletons were not clad— a cost-cutting casualty of the Great Depression— but the steel-frame towers, never intended to be visible, are among the most distinctive features of the New York cityscape. This 1930 photograph shows construction of the south ramp from Riverside Drive. When completed, the George Washington Bridge contained the world's longest suspension span, stretching 3,500 feet between the two towers and suspended 200 feet above the Hudson River. It originally had only a single 90-foot-wide roadway with two 10-foot sidewalks. A lower deck, begun in 1958 and completed in 1962, was originally planned for four electric railroad tracks but was changed to accommodate additional lanes of automobile traffic.

1927–41

BLACK HILLS, SOUTH DAKOTA

Midway between the East and West coasts four 60-foot-high faces are carved into a stone mountain 500 feet above the ground. Mount Rushmore National Monument would, the sculptor Gutzon Borglum (1867–1941) declared, immortalize America's "empire builders," the presidents most responsible for the "creation, preservation and expansion of the Union"— George Washington, Thomas Jefferson, Abraham Lincoln, and Theodore Roosevelt. Borglum, who also recreated the lantern of the Statue of Liberty in 1916 and initiated work on the Confederate generals on Stone Mountain, Georgia, in 1923, believed that "a monument's dimensions should be determined by the importance to civilization of the event commemorated." Accordingly, the faces were carved to the scale of men 465 feet tall; from chin to top each head is about 60 feet, each nose is 20 feet long, each mouth is 18 feet wide, and the eyes are 11 feet across. The faces are so large that their features were first blasted with dynamite and drilled with jack hammers and then carved. Originally promoted by South Dakota chambers of commerce, the monument, begun in 1927 and completed in 1941, eventually cost $1 million, 85 percent of which was funded by congressional appropriation. Borglum died a few months before completion of the work, which was then supervised by his son, Lincoln Borglum. The faces on the mountain are visible for 60 miles.

109

LOUISVILLE PALACE THEATER

LOUISVILLE, KENTUCKY

Many movie palaces of the 1920s were called atmospheric theaters, and the atmospheres they evoked were exotic and far away: Egyptian temples, Italian courtyards, Persian gardens, Mediterranean villas, and Spanish patios. They were affordable fantasies for anyone who could pay the price of a ticket. One of the greatest designers of atmospheric theaters was John Eberson (1875–1954), an Austrian immigrant who regarded the ornate, eclectic picture palaces of the 1920s as places where "our fancy is free to conjure endless tales of romance"—the essence of the silent cinema. One mark of Eberson's dramatic genius was his creation of stars and clouds filling indoor skies. He covered ceiling domes in blue plaster and added drifting clouds and twinkling star lights, an illusion produced by a Brenograph magic-lantern machine in which a strip of negatives was passed in front of a 1,500-watt light bulb. In its appropriately named Palace, Louisville received the finest of Eberson's Spanish courtyards—its side-wall villas and tile-roofed proscenium conjuring up nights in Iberia. In the grand foyer of the theater, first a Loew's and then a United Artists property, a barrel-vaulted ceiling presents portraits in relief. Walls are lined with medallions of historic figures, including one of Eberson himself. Unlike many of the grand picture palaces, this one remains.

BUI

HOOVER DAM

1931–35

BOULDER CITY, NEVADA

Southern California's thirst for water gave birth to Hoover Dam. Dams are among the world's greatest engineering feats, and Hoover Dam, originally known as the Boulder Canyon Reclamation Project, is perhaps the greatest of America's dams. Everything was at a huge scale. Construction, begun in 1931, required the largest federal contract to date: $31 million. The dam rises 726 feet above its foundations, is 660 feet thick at its deepest elevation, and is 1,244 feet long at the top of its Moderne crest, the work of Los Angeles architect Gordon Kaufmann (1888–1949); John Lucian Savage of the Bureau of Reclamation supervised the construction. Cofferdams were sunk to bedrock to allow dry excavation for the foundation and then the Colorado River was diverted into tunnels that later became spillways. Thirty-foot-diameter steel penstock pipe was installed by cableway for placement in the upper Nevada header tunnel. The photograph shows a bend section for installation below one of the intake towers being lowered over the rim of the canyon; a worker sitting in the pipe gives an idea of its tremendous size. To counteract the heat produced by the concrete as it hardened, 582 miles of one-inch water pipe was embedded throughout the structure to cool the concrete as it set so that it would cure properly. More than 3.25 million cubic yards of concrete eventually was placed by an around-the-clock work crew housed in a nearby town known as Boulder City.

ARTHURDALE

1933–34

REEDSVILLE, WEST VIRGINIA

The Arthurdale Subsistence Homestead Project,
championed by First Lady Eleanor Roosevelt
as a noble experiment during the New Deal, was
one of several such government projects during
the Great Depression that attempted to provide
low-cost housing, skills training, and other
services to alleviate poverty and joblessness.
The federal government bought Richard Arthur's
1,100-acre farm here and established a new
community; it was designed to provide cooperative opportunities for small-scale farming and
craft manufacturing of furniture and wrought-
iron articles as a supplement to the local
vacuum cleaner and shirt factories in Reedsville.
In November 1933 the government ordered and
rushed to Arthurdale 50 prefabricated Cape
Cod–style residences in the hope that some
families could be housed by Christmas. The
buildings, however, were inadequate for the
harsh winters and had to be rebuilt. The
waiting homesteaders themselves insulated and
plastered the houses, corrected the entire
plumbing system, and remodeled the community hall. On opening day, June 6, 1934, 38
families moved in. The families paid $5,000 for
each house, but the actual cost, with the
rebuilding, was about twice as much. Eventually
165 families would be housed at Arthurdale.
Noted photographer Walker Evans (1903–75),
during his work for the Farm Security Administration, captured additional houses being built
in June 1935. Arthurdale succeeded in raising
living standards for homesteaders but, because
of poor economic planning and the inability to
attract new industry, failed to be a self-
sufficient cooperative and declined after 1946.

GOLDEN GATE BRIDGE

1933–37

SAN FRANCISCO–MARIN COUNTY, CALIFORNIA

Hundreds of thousands of San Francisco Bay Area revelers celebrated the fiftieth anniversary of the Golden Gate Bridge on a Sunday morning in May 1987 by walking across it, creating a jam-packed mass of humanity in the middle. The 250,000 people crowded onto the roadway created significantly greater stress on the bridge than the usual traffic, causing the deck to deflect 7 to 12 inches and creating grave concern among bridge authorities. The Golden Gate Bridge withstood the excesses of its birthday celebration without harm, while demonstrating that any weakness or flaw in the engineering would have resulted in a San Francisco disaster equal to the city's 1906 earthquake. Joseph B. Strauss (1870–1938) designed the bridge to carry a load of 160 million pounds on its clear span of 4,260 feet over the water. Such a bold suspension bridge could not have been built or funded before World War I. As much as residents wanted to cross the Golden Gate by car, many were convinced that the bridge would "certainly mar if not utterly destroy the natural charm of the harbor famed throughout the world," as *The Wasp*, a local publication, whined in 1925. Execution of the red landmark took two decades. The actual work took four years, culminating in towers 746 feet tall, two cables with a diameter of more than 36 inches, a roadway 220 feet above the water, and an overall length exceeding 8,900 feet.

GRAND COULEE DAM

1934 – 42

COULEE DAM, WASHINGTON

Measured in cubic yardage of concrete, the Grand Coulee Dam—a 550-foot-high, 4,173-foot-long concrete gravity dam—is the largest structure in the Western Hemisphere and among the world's most massive structures. Franklin Roosevelt saw this project on the Columbia River, the second largest in the United States, as an ideal means of bringing to the Pacific Northwest his vision of economic revitalization through public works. The federal government authorized the project in 1933, and in the summer of 1934 Roosevelt presided over the ceremonies marking the start of construction. Water stored behind the dam serves two purposes: much of it is released through the dam's hydroelectric plant and is used to generate vast quantities of power, while some of it is used to pump water 280 feet to the Grand Coulee proper, forming a secondary reservoir that supplies irrigation water to the Big Ben region, an area of several hundred square miles in central Washington. Because of a 1,000-foot difference in elevation between the gorge and the area requiring irrigation, a system of canals and tunnels was built. Unlike the dams of the Tennessee Valley, the Grand Coulee was originally conceived to reclaim arid lands for farming, while the first TVA dams actually flooded existing farmland to create a fall of water for generating power. During World War II the dam proved critically important to activities at the Hanford Reservation, where much work on the first atomic bombs took place.

GREENBELT

1935–37

GREENBELT, MARYLAND

When the Suburban Division of the Resettlement
Administration finally received funding to build
its proposed greenbelt towns, the planners were
told to begin with projects in Maryland, Ohio,
and Wisconsin. As model planned communities,
the greenbelt towns were conceived to
demonstrate garden-city planning principles,
offer good housing for families of modest income,
and put the unemployed to work. Planning and
design for these new towns had taken place at
an inappropriately lavish 54-room mansion on
Massachusetts Avenue in Washington, D.C., the
former residence of Evelyn Walsh McLean. In
1935, looking for quick results, President Franklin
Roosevelt directed that land acquisition and
construction begin by the end of the year and
that the project be completed in six months.
Because land had already been acquired in
Maryland, 800 transient laborers were hired two
days after the presidential order was signed. No
construction plans or materials were on site
when the workers arrived, so some were issued
shovels to begin digging a lake. Eventually 200
workers spent a year turning a swamp at
Greenbelt's entrance into a 23-acre lake. The
Greenbelt town plan contained a number of
superblocks of 14 to 18 acres, each with 120
houses. These were surrounded and protected
by wooded areas and served by walkways
leading to a prototype suburban shopping mall
built in a streamlined Art Deco style; the
photograph shows the bus terminal and theater
under construction. Greenbelt, Maryland, was
completed according to plan, although a year
behind schedule, and was followed by Greenhills
near Cincinnati and Greendale near Milwaukee.

ROCKEFELLER CENTER ICE RINK

1936

NEW YORK CITY

When the lower plaza in front of the RCA
Building at Rockefeller Center was opened in
1934, it was flanked by shops. They were off the
beaten track, however, and failed to attract
customers. To animate the space, they were
replaced with an ice-skating rink and a
restaurant, both of which opened on Christmas
Day 1936. As the photograph shows, a grid of
water pipes lies under the rink's surface. The
bronze-gilded figure *Prometheus* by Paul
Manship (1885–1966), located against the west
wall of the rink, was installed in 1934. One of
Rockefeller Center's well-known traditions is its
annual community Christmas tree—a ritual that
began when workers on the construction site set
up a tree on December 24, 1931. The tree has
reappeared each year on the same spot, now the
approach leading to the ice rink.

RINGLING BROTHERS AND BARNUM AND BAILEY CIRCUS

BROOKLYN, NEW YORK, AND NEWARK, NEW JERSEY

The American circus has always been presented both inside, in indoor arenas, and outside, beneath the open sky and under tents. At tented circuses, caged animals and other auxiliary curiosities performed under a Big Top. From 1927 to 1956 the Ringling Brothers and Barnum and Bailey Circus traveled the country with its big tent. A typical program was produced on two stages and three rings and included a grand entry, five high-wire acts, 25 performing elephants, six aerial acts, contortionists, balancing and juggling acts, a monkey-climbing act, dogs, pigeons, clowns, seven acrobatic acts, five groups of trained sea lions, a lion-taming act, eight trapeze acts, and one number featuring 68 horses and Shetland ponies, four zebras, and five camels, all in one ring. The logistics of moving such a circus were impressively complex. During a typical 1930s season 525 horses traveled with the show, including 200 ring stock and 325 draft animals that helped move baggage and equipment from rail depot to circus lot and helped erect the tent. By the late 1930s Ringling Brothers substituted first elephants and then tractors for horses and began to move the whole ensemble by truck rather than rail, significantly reducing the number of roustabouts, wranglers, and draft animals. By 1948 a large mechanical apparatus attached to a truck was used for loading, transporting, and unloading canvas, although dozens of men and animals working 6 to 10 hours were still required to raise the Big Top.

FALLINGWATER

1936–37

BEAR RUN, PENNSYLVANIA

During construction of the Pennsylvania country
retreat for Edgar J. and Liliene Kaufmann, a
continuing dialogue about the stability of the
dramatic cantilevers and parapets ensued
between architect and client, architect and
engineer, architect and contractor, contractor
and engineer, and contractor and client. Edgar J.
Kaufmann, Jr., who coordinated the construction
of Fallingwater, believed that there were many
mistakes, disagreements, and oversights in the
calculation of the amount of steel and the tilt of
the forms as the building went up. But for
Kaufmann these faults, which are evident in
cracks in the concrete, do not impugn the ability
of the architect, Frank Lloyd Wright (1867–1959):
"Comparable situations indicate an answer,
bearing in mind that the architect and his client
knew the design of Fallingwater was an
exploration beyond the limits of conventional
practice. The small deflections (up and down)
were not foreseen; neither were those (from side
to side) in early skyscrapers, yet these are now
accepted as normal. Some of the great
monuments of architecture have suffered
structural troubles precisely because they were
striving beyond normal limitations."

JOHNSON WAX
ADMINISTRATION BUILDING

1936–39

RACINE, WISCONSIN

When Frank Lloyd Wright presented his plans for the S. C. Johnson offices and laboratory, the company's building department immediately challenged his concept. Columns that appeared to be supporting the roof were wide at the top and tapered at the floor—the reverse of the basic structural principle that base size is proportional to the load carried. To prove otherwise, Wright constructed a prototype column and loaded it with sand bags weighing 12 times the load it was calculated to carry. The column did not waver, and construction proceeded. The 54 dendriform columns, hollow structures of poured concrete over wire mesh,

were each topped with an 18-foot lily-pad top. From its structural system to the office furniture to the use of Pyrex glass tubing to permit natural lighting into its large, open central space, the building was innovative. "We became a different company the day the building opened," declared Samuel C. Johnson, chief executive officer and chairman of Johnson Wax. "We achieved international attention because that building represented and symbolized the quality of everything we did in terms of products, people, the working environment within the building, the community relations and—most important— our ability to recruit creative people."

SHARECROPPER'S CABIN

BUTLER COUNTY, MISSOURI

As progress has homogenized and eliminated vernacular regional forms of building, small houses for tenant farmers that dotted the rural landscape of the South for more than a century have only recently disappeared. The share-cropper's cabin grew out of slave dwellings built in the first half of the nineteenth century—small, minimal living places often no larger than 10 feet square. Frederick Law Olmsted, who toured the South in the late 1850s, observed that most country dwellings were made of rough-hewn logs; windows were simply openings with cloth or skin coverings, and the floor, elevated on four corner posts, was two or three feet off the ground. These traditional buildings may have derived from dwellings in the bush of West and Central Africa. Knowledge of their construction probably migrated with the slaves in the eighteenth century. Often these houses had no opening other than a single doorway and featured front porches where domestic chores were generally performed. The main differences between slave quarters and the later sharecropper's cabins occupied by either white or black tenant farmers were the size of the dwelling—the houses of free laborers were often more than twice as large—and their proximity to other houses. Slave dwellings were usually built in "quarters" or "on the street," not far from the overseer or the master's big house. For tenant farmers, who lived out in their rented fields, domestic life was solitary; the nearest house was often miles away.

144

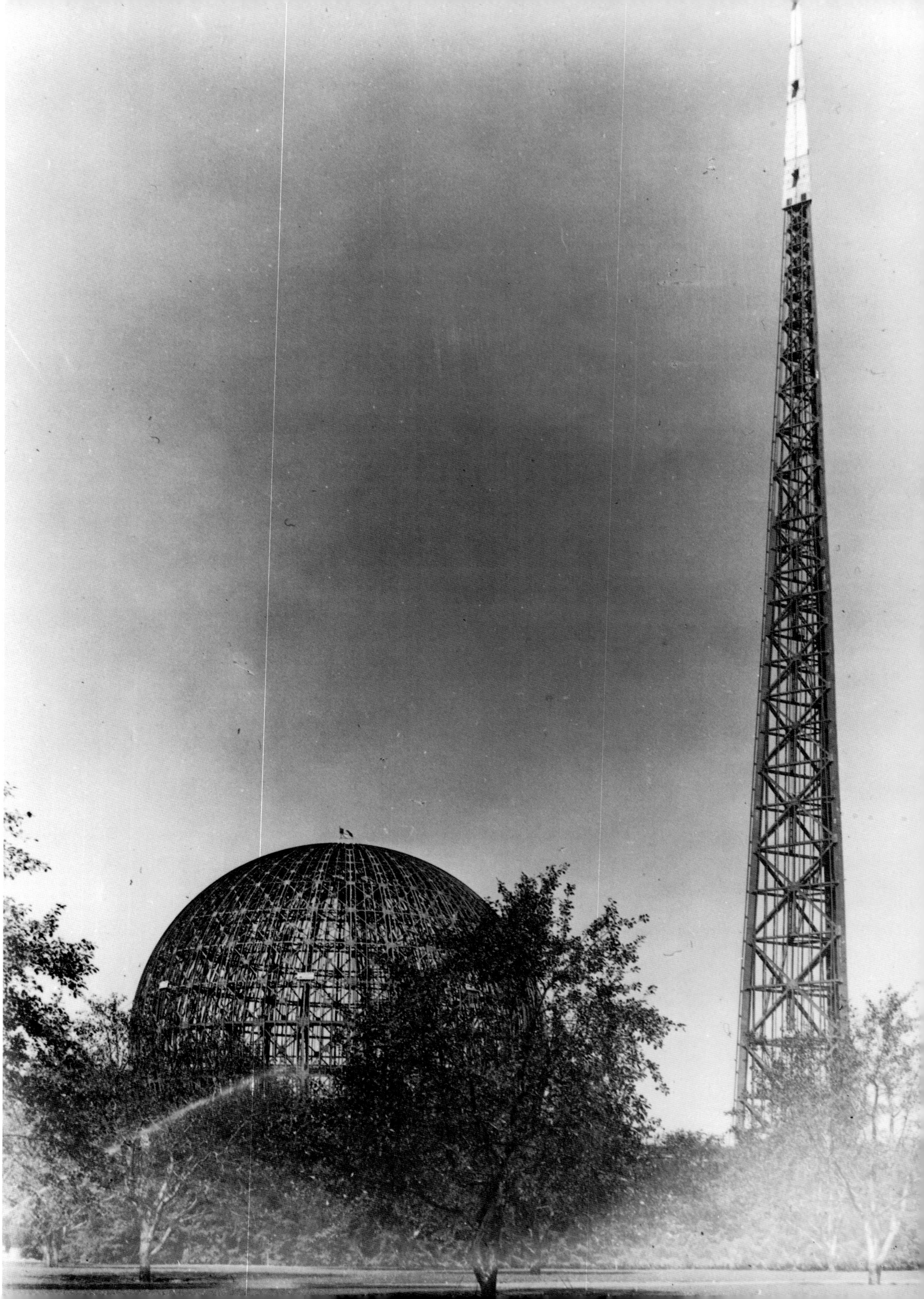

TRYLON AND PERISPHERE

1939

FLUSHING MEADOW, NEW YORK

The 1939 New York World's Fair, the most ambitious international exposition since the first New York's World's Fair in 1853, attempted to give a glimpse of the World of Tomorrow. The fair's 200 buildings represented the pinnacle of American modern design. Clearance of the 1,216½–acre site, originally a huge garbage dump on a marshy bog, was the largest land reclamation project in the United States to date. The fair's thematic and architectural centerpiece was the Trylon and Perisphere, created by Wallace K. Harrison (1895–1981), who later designed the United Nations Building, and J. André Fouilhoux (1879–1945) to "express the civilization of the future with confidence." The Trylon was a 610-foot-high obelisk; the Perisphere, a hollow ball 180 feet in diameter and 18 stories high, was the largest globe ever built. The two icons of the fair were constructed of 2,000 cubic yards of concrete and reinforced steel and 3,000 tons of structural steel resting on more than 1,000 pilings of Douglas fir creosoted for durability. The total weight of the structures and foundations was 10,000 tons. Visitors rode part of the way up the Trylon on what was then the world's highest escalator and then entered the Perisphere, stepping onto one of two moving rings from which they viewed the vast diorama of Democracity, a planned city complex of the future designed by Henry Dreyfuss (1904–72). After a short film, visitors walked back to ground level on a ramp called the Helicline.

SET OF MR. SMITH GOES TO WASHINGTON

1939

HOLLYWOOD, CALIFORNIA

Director Frank Capra began work on *Mr. Smith* by organizing a scouting trip to Washington to shoot backgrounds, photograph monuments and buildings and hundreds of small details in the Capitol, and locate a technical adviser for the film. Capra's objective was to create an authentic reproduction of the Senate chamber in a Columbia studio sound stage. According to Capra, Lionel Banks, the film's art director, recreated in "100 days … what had taken 100 years to build." In addition, exact duplicates of a cloakroom, committee room, press club, and hotel suite as well as close-up details of monuments had to be researched and constructed. At the same time Capra personally interviewed players for the movie's 186 speaking parts. Unlike most movie sets, the Senate chamber was constructed as a four-sided well; the boxed-in space enclosing the floor of the Senate was surrounded by gallery seats backed by a high wall with niches for statues of 20 former vice presidents. Scenes were photographed on three levels—the floor, the rostrum, and the gallery—with story action sometimes taking place on all three levels at once. Rather than use a single camera—moving it with its platform, sound booms, and lights for each new set-up—the crew planned multiple camera positions and a multiple sound channel method to capture a large number of scenes without having to move the equipment.

JEFFERSON MEMORIAL

1939 – 42

WASHINGTON, D.C.

When the Jefferson Memorial was dedicated on April 13, 1943, marking the bicentennial of the third president's birth, the 19-foot portrait sculpture inside was cast in plaster. Because of a wartime restriction on metals, a permanent bronze casting was not commissioned until 1946. The building, designed by John Russell Pope (1874 – 1937), was inspired by Jefferson's design for the Rotunda at the University of Virginia, which itself was adapted from the Pantheon in Rome. Inscriptions on four panels, each with quotations from Jefferson's writings, surround the statue. Thirty years after the Jefferson Memorial was completed, Frank W. Fetter, professor at Northwestern University, observed that the quotations contained inaccuracies. In fact, the architects deliberately made several small changes in wording, spelling, and punctuation so that the typography of the inscriptions would be perfectly spaced and aligned on the panels. The Jefferson Memorial overlooks the Tidal Basin, ringed by its famous Japanese cherry trees.

151

ADOBE RESTORATION

1940

CHAMISAL, NEW MEXICO

Traditional adobe houses in the southwestern United States were made by both men and women, although only men performed the heavy lifting. The adobe bricks — 10 by 18 by 5 inches and weighing 50 pounds each — were made of a mixture of clay, sand, and straw and then poured into wooden molds. The bricks were dried in the molds, after which they were laid on stone foundations to form walls. To hold the adobe bricks in place, thick mud was used between the layers. On top of the walls, wooden beams called vigas were placed across the width of the house and then covered with brush, adobe, and several inches of dirt. The walls were finished by women who troweled a mixture of red, brown, or white plaster over the bricks, inside and out. The exterior was usually replastered every year or two, a process caught by Russell Lee (1903 – 86) in this photograph taken for the Farm Security Administration.

NAZI PARTY HEADQUARTERS

ca. 1941

MUNICH, GERMANY

At an exhibition of the work of artist Martin Mächler in 1927 Adolph Hitler was introduced to the idea of a grandiose Berlin, its streets conceived on the scale of those in Paris, its public buildings designed to accommodate great throngs of people. After the 1936 Berlin Olympics Hitler directed Albert Speer, mayor of Berlin and later the official Third Reich architect, to create for the federal capital a grand avenue, 130 yards wide, with a public assembly hall and a triumphal arch at each end. The meeting hall was to be large enough to enclose a crowd of 150,000 cheering citizens; the arch was to rise to 400 feet, more than three times the height of the Arc de Triomphe. With Speer working on a grand plan for Berlin, Hitler selected a dozen other architects to develop similar plans for other German cities, including Munich. But as World War II progressed, there was no time and few resources to execute these plans, and existing structures, such as the Nazi Party headquarters, were adapted to wartime requirements. Rather than create new structures, the Third Reich simply renovated existing neoclassical buildings and added a few signature touches, such as swastikas. Writing 30 years later, Speer belatedly commented, "It strikes me as rather sinister that in the midst of peacetime, while continually proclaiming his desire for international reconciliation, he was planning buildings expressive of an imperial glory which could be won only by war." Hitler used slave labor to build roads, buildings, and other public works during the war, none of which approached his megalomaniacal vision for a new Germany.

1942

INDEPENDENCE, CALIFORNIA

During World War II more than 110,000 people of Japanese ancestry in the United States were removed from their homes and confined in a number of relocation centers. The native-born Japanese and their children were then evacuated to 10 War Relocation Authority camps in California, Arizona, Idaho, Utah, Colorado, and Arkansas. The largest of the centers, Manzanar, in California, was 560 acres. Nine wards of four blocks each were constructed, with each block containing 16 one-story buildings, 20 by 100 feet. These barracks were divided into apartments of various sizes, where as many as eight people were assigned to a space 20 by 25 feet. The amount allotted for the construction of barracks, a mess hall, a recreation hall, a communal kitchen, and bathhouses for each block was $376 per person. The whole area was surrounded by barbed wire and guard towers, with another outer perimeter containing 5,700 acres, of which 1,500 acres were farmed by the internees assigned to agricultural work. Each barracks apartment house was constructed of six-inch boards on a wood frame, covered with tar paper secured by battens. As photographer Russell Lee noted for the Office of War Information, these structures were "theater of operations" housing intended for combat-trained soldiers to be used on a temporary basis. They were equipped only with iron cots covered by straw mattresses and army blankets, plus a small oil heater. At the relocation centers, men, women, and children were confined in freezing cold and extreme heat for up to three and a half years.

THE PO

STWAR ERA

LEVITTOWN

1947–51

LONG ISLAND, NEW YORK

After World War II American GIs married and began having children in record numbers, creating an enormous demand for housing. A new type of house that most newlyweds could afford was built by Abraham Levitt. In fact, a veteran with a government loan could buy one of Levitt and Sons' houses on Long Island for just $56 a month. Levitt, the largest builder on the East Coast, revolutionized the concept of residential construction, using economies of scale and standard techniques—much like the auto industry—to achieve great efficiencies. He produced his own lumber and nails, warehoused the lumber precut to size, and preassembled in factories the staircases, kitchen cabinets, and plumbing fixtures needed for his small houses. Using poured concrete slabs for foundations and prefabricated walls, he was able to create almost identical 800-square-foot houses that sold for less than $8,000 in 1947. All Levitt houses had the same basic floor plan, no basement, and few options. Construction was broken down into 26 steps performed by specially trained crews that moved from house to house, each attending to a specific task; shown are workers leveling the ground in a row of ranch houses. By 1950 Levittown consisted of more than 15,000 houses and had a population of 60,000—for a time the largest community built by a single developer. Other Levittowns were built in Bucks County, Pennsylvania, and Burlington County, New Jersey.

1948-52

WASHINGTON, D.C.

When Franklin Roosevelt's family moved out of the White House in 1945 with 13 truckloads of furnishings after a 15-year residency, numerous cracks were discovered that eventually revealed serious defects in the building's structure. Renovation, it was generally agreed, was necessary. Heading up the project were Maj. Gen. Glenn E. Edgerton, executive director of the Commission for the Renovation of the Executive Mansion; Col. Douglas H. Gillette, assistant executive director of the commission and an engineer; Lorenzo S. Winslow, architect of the White House; and Harbin S. Chandler, Jr., assistant to the architect and chief designer. Work on the 156-year-old Executive Mansion, which began in 1948, was not a preservation project in any sense. The interior was completely gutted, and the original fabric—moldings, columns, doors, window frames and hardware—was mostly consigned to the junk heap. The building, whose shallow footings rested on an insecure clay base, had to be supported on temporary steel plates as 126 new concrete piers along the perimeter of the structure were poured. In the process a new basement and subbasement were dug. All that was left of the original building was the stone walls, and President Truman refused to allow a doorway to be cut wider as passage for a bulldozer and several dump trucks. So the machinery needed to dig the new basement had to be dismantled, brought into the White House shell in parts, and reassembled there. A new steel inner frame was constructed inside the rectangle of the original walls.

FONTAINEBLEAU HOTEL

1952-54

MIAMI BEACH, FLORIDA

Fantasies spun by the movies in the 1930s
were sometimes transformed into architecture
in the 1950s, as Americans who could only
dream of a vacation to the tropics during the
Great Depression could actually afford to make
the trip during the postwar boom. Inspired by
Florenz Ziegfeld and Busby Berkeley, movie
palaces of the 1920s, and flourishes from the
French Renaissance, architect Morris Lapidus
(1902–) fashioned a landmark of popular taste
that created a whole new style of resort—an
eclectic mix of styles that completely rejected the
austerity of other large buildings of the era. The
oceanside Fontainebleau in Miami Beach, whose
theatrical interiors used variations in floor levels
to dramatize the space, displayed an ornamental
richness in a contemporary atmosphere. The
Mediterranean-style Harvey Firestone mansion,
originally part of the three-mile-long Millionaires
Row developed in the 1930s, was torn down
to make way for the Fontainebleau's grassy
esplanade. Lapidus, designer of some six dozen
hotels, was dubbed the High Priest of High
Kitsch by architecture critic Ada Louise Huxtable
in the *New York Times:* "His work is often
wonderfully, pratfall funny—these are the best
esthetic sight gags in the world...."

1954-55

ANAHEIM, CALIFORNIA

Created by Walt Disney to fill what he perceived as a lack of interesting yet wholesome entertainment for Americans, Disneyland mixed amusement park rides with world's fair–type exhibitions, all centered around multiple themes: turn-of-the-century Main Street, U.S.A.; Fantasyland, populated with Disney cartoon characters; the African Adventureland; the American West in Frontierland; and what passed for the future in the 1950s, Tomorrowland. Because the idea was so novel, he had difficulty finding investors for the $17 million project (originally estimated to be $4 million); in return for $5 million, ABC received a 35 percent interest in Disneyland. Construction began in 1954 with the bulldozing of the orange and walnut groves on the site and continued with the reinvention of nature and the world inside the park. Disneyland represents the mass merchandising of what French theorist Louis Marin called the "imaginary relationship that the dominant groups of U.S. society have with their history." The 180-acre site with its 146-foot replica of the Matterhorn has attracted nearly 200 million visitors over the past four decades. Its immense popularity has spawned East Coast and European counterparts—Disney World (1969–71) in Orlando, Florida, and Euro Disneyland (1989–92), near Paris—and a proposed Disney America, in Prince William County, Virginia.

SYDNEY OPERA HOUSE

1956-68

SYDNEY, AUSTRALIA

The Sydney Opera House has evolved into
a majestic icon of the mid-twentieth century:
a combination of massive ambition and limited
imagination that reflected global political
developments of the era. Danish architect Jørn
Utzon (1918–) provided the winning design in
the 1956 international competition; one of the
four jury members was the architect Eero
Saarinen. Utzon's evocative image of billowing,
sail-like forms may have influenced Saarinen,
who was at work on the TWA Terminal at
the time. A number of prefabricated concrete
shells with ceramic facing (some matte, some
shiny), designed for their acoustical benefits
rather than their appearance, dominate the
building. A 2,700-seat concert hall occupies the
area beneath the largest series of shells, with
a smaller 1,500-seat opera house adjacent and
two miniature shells on the street side of the
building. Cost overruns and difficulty in constructing the roof forms resulted in a gulf between
ambition and execution.

TWA TERMINAL

1956–62

NEW YORK CITY

The design for the TWA Terminal at John F. Kennedy International Airport (originally Idlewild Airport) has been both praised and damned by critics. Vincent Scully called it "Pseudo-concrete choked with steel ... a whammo structural exhibition which is always threatening, visually at least, to come apart at the seams." Edgar J. Kaufmann, Jr., considered it "a festival of ordered movements and exhilarating spaces," while Eero Saarinen (1910–61), the building's architect, said simply that he wished "to catch the excitement of the trip." His goal was "to design a building in which architecture itself would express the drama and specialness and excitement of travel ... a place of movement and transition.... The shapes were deliberately chosen in order to emphasize an upward-soaring quality of line." The structure appears to be only a roof, a shell-like enclosure divided into four parts, each separated by a band of skylights. All is held up by four massive Y-shaped columns, an engineering feat accomplished by Ammann and Whitney. AIA medalist Balthazar Korab, a former associate of Saarinen's, photographed both the TWA Terminal and Saarinen's other airport, Dulles.

DULLES INTERNATIONAL AIRPORT

1958 – 62

CHANTILLY, VIRGINIA

Ushering in the jet age, Eero Saarinen's Dulles
International Airport was described by Ada
Louise Huxtable as "a building of singular beauty
... the kind of architecture of a force and
importance unparalleled since the unique
synthesis of philosophy and style that produced
the Renaissance." In this design Saarinen pushed
the technological possibilities of molded concrete
forms: a sequence of reinforced concrete roof
supports appears to lift from their stage, an
extraordinary illusion—as if the whole 600-foot-
long structure is about to take flight. Light
suspension cables support the roof, and the
concrete piers, 65 feet high in the front and 40
feet in the back, slope outward to counteract the
pull of the cables—a tension caught in Balthazar
Korab's photograph. "But we exaggerated and
dramatized the outward slope," Saarinen noted,
"to give the colonnade a dynamic and soaring
look as well as a stately and dignified one."
He believed it the best design of his career, a
view corroborated by many. The terminal, named
for Secretary of State John Foster Dulles, is
located on a 10,000-acre flat site 25 miles from
Washington, D.C. Saarinen's design, one of the
few under-50-year-old buildings declared
eligible for the National Register of Historic
Places, is now being expanded.

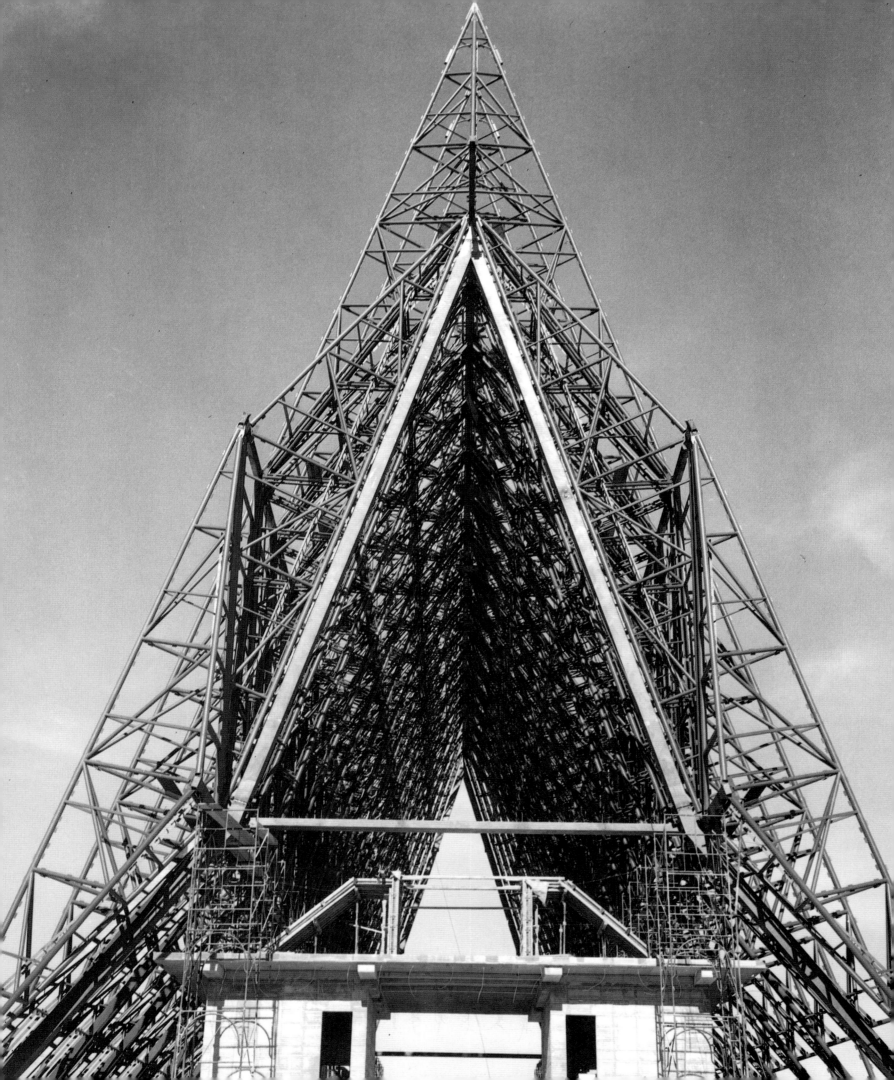

1958–62

COLORADO SPRINGS, COLORADO

Severely and dramatically modern, the design for the U.S. Air Force Academy was radical for a military facility. The chapel, located at the foot of the Rampart Range of the Rocky Mountains, is the architectural centerpiece of the Skidmore, Owings and Merrill–designed campus. An academy football coach has been quoted as saying, "We don't know whether to pray in it, for it, or at it." The nontraditional design of the campus, particularly the chapel, ignited a public outcry and led Congress to hold hearings on the issue. Frank Lloyd Wright, who briefly headed the architectural group that was one of two finalists for the project, was called to testify before the House Appropriations Committee. He denounced the design as a "factory for birdmen," urged that a new competition be held, and, temporarily abandoning his lifelong antimilitary stance, offered to design the chapel for free. The committee did not accept his suggestions but did recommend that funding for the chapel be withheld indefinitely, citing concern that the proposal did not reflect the best traditions of American architecture. SOM slightly modified the design, and Congress restored funds for the chapel in 1958, two years after funding for the other buildings was approved. The only academy building open to the public, the chapel includes sanctuaries for Protestant, Catholic, and Jewish services. Structurally, it is a series of triangulated tetrahedrons made of steel tubing with aluminum facings; one-foot bands of stained glass lie at their junctures. Construction of the entire campus was documented in detail by Stewarts Photographers of Colorado Springs.

1959-65

ST. LOUIS, MISSOURI

Eero Saarinen's Gateway Arch is a tribute to Thomas Jefferson, who designed the graceful archways of the University of Virginia and who as president masterminded the Louisiana Purchase, the cornerstone of America's westward expansion—of which St. Louis was the gateway. Winner of the 1947–48 design competition for the Jefferson National Expansion Memorial, Saarinen garnered the $50,000 grand prize from among 172 entrants. His design was realized through the expertise of the structural engineers Severud-Elstad-Krueger Associates. Eighty-four acres of historic buildings on the St. Louis riverfront were razed for the memorial's site, leaving only the old courthouse. The arch, the nation's tallest monument, is 630 feet high with a span of the same length and required 900 tons of stainless steel for its cladding.

Construction above 72 feet proceeded with the use of two "creeper" derricks mounted on tracks fastened to the arch. These rigs climbed to predetermined stations, and then the derricks lifted arch sections into place. The photograph below shows the last section being installed on October 28, 1965, when the derricks were positioned at 610 feet. An 80-ton trussed stabilizing strut had been installed between the two legs at 530 feet for support until the legs were joined in the complete arch. Jacks mounted on top opened to allow the last section to be inserted; when the jack pressure was released, the natural inward thrust of the legs clamped the section in place. After it was connected, the derricks were lowered, tracking was removed, bolt holes were filled, and the stainless steel surface was polished.

1961-63

CHICAGO

Marina City, twin 60-story cylindrical towers, is a tribute to mixed use in a congested downtown where the tax burden of a single-use building would be too great. Writing in *Architectural Record*, Bertrand Goldberg (1913–), the project's architect, explained: "We cannot burden business buildings used 35 hours a week or apartment buildings used at night and over weekends with our total tax loads. We can no longer subsidize the single shift use of our expensive city utilities." Marina City combined commerce, recreation, and education uses on lower floors with residences above, an elaboration of a scheme introduced to Chicago by Louis H. Sullivan (1856–1924) in his Auditorium Building (1889), which incorporated offices, a hotel, and a

theater under one roof. Goldberg, once a student of Mies van der Rohe's, incorporated the principle of structural ethics in his buildings but found more freedom in reinforced concrete than in steel. In Chicago three of his most important designs make use of curved forms in concrete— Marina City, the Raymond Hilliard Housing (1964–66), and the Prentice Women's Hospital (1970–75). In Marina City the lower third of each tower contains helical parking ramps, and the upper two-thirds consists of 40 floors of apartments, divided into 16 petal-like segments. The ramp and apartment floors are cantilevered from a central core with an outer ring of secondary supports; on the apartment floors curved balconies project from each segment.

ASTRODOME

1963-65

HOUSTON, TEXAS

With typical Texas hype, Judge Roy Hofheinz built the Astrodome largely with public funds, calling it the eighth wonder of the world. Designed by the firms of Hermon Lloyd and W. B. Morgan and Wilson, Morris, Crain and Anderson, it was the first permanently enclosed air-conditioned sports arena built to accommodate both baseball and football games. With its 642-foot span, the 218-foot-high trussed roof structure can accommodate as many as 66,000 patrons for an entertainment or sporting event. The Astrodome was the first arena to use Astroturf, an artificial grass developed in 1964 that became the preferred playing surface as other large indoor stadiums were built. In 1989 the owner, Harris County, demolished its 474-foot-long, four-story electrified scoreboard to make room for more seating. Sited on a 260-acre tract that includes a 30,000-car parking area, the Astrodome has evoked varied responses. Vicky Alliata concluded in 1974 that "the whole thing far surpasses all current definitions of kitsch, obscenity, and bad taste," but other critics note that being able to play ball when it's raining more than compensates for any design flaws.

JOHN HANCOCK CENTER

1966–70

CHICAGO

Bruce Graham (1925–) of Skidmore, Owings and Merrill, architect of the Hancock Center and later the Sears Tower, employed an early and extensive use of computer-assisted design (CAD) techniques in both the structural engineering and the design of the skyscraper. With the assistance of Fazlur Khan (1929–), Graham exposed diagonal bracing across a tapered skeleton to provide a trussed box to withstand high winds from Lake Michigan whipping against its 1,107-foot height. Diagonal bracing is the way in which designers solve the structural problem of stabilizing very tall buildings against wind loads. Usually hidden, here they provide a graceful design element as well as structural support. The Hancock Center mixes tenants in the best Chicago tradition: retail stores occupy the first 5 floors; parking for 1,200 cars takes up floors 6 through 12; offices and mechanical services, floors 13 through 17; commercial office space, floors 18 through 41; and mechanical equipment, floors 42 and 43. A sky lobby, on floors 44 and 45, is where apartment dwellers change elevators, shop, swim, and dine. Floors 46 through 92 contain 705 condominium apartments; an observation lounge and a restaurant are located on the next 3 floors, and mechanical and television equipment fill the remainder of the space through floor 100.

U.S. PAVILION, EXPO 67

1967

MONTREAL, QUEBEC, CANADA

The skyline of Expo 67 was distinguished by two architectural icons: Buckminster Fuller's geodesic dome housing the U.S. pavilion and Moshe Safdie's Habitat 67, a stair-stepped prefabricated concrete apartment house at the Cité du Havre, an all-in-one-building village along the riverbank. Safdie and Fuller presented differing expressions of the same ideas: that industrialization and mass production would go far to cure housing ills and that expertise and a good system are all that is needed to solve the world's problems. Throughout his life Fuller (1895–1983) had promoted his geodesic designs as a way to solve many problems—energy conservation, environmental control, comfortable living conditions. His Expo 67 dome, a three-quarters sphere, was as tall as a 20-story building but was fabricated from a complex set of mathematical tables rather than conventional drawings. These figures provided workers with exact specifications for stamping out the components assembled to create the dome's frame. Fuller intended that the lightweight aluminum struts be bolted together so that the structure could later be moved and reused, but fair officials insisted that they be permanently welded in place. As a result, construction was much slower (the dome was still being worked on when the fair opened), and after the fair closed the city had a white elephant on its hands. Later when workers were rewelding problem areas, sparks ignited a panel of the Plexiglas windows that served as the skin, and within minutes fire had destroyed the dome's steel skeleton.

184

SEARS TOWER

1970-74

CHICAGO

The Sears Tower inspired a completely different
solution for the Windy City, even though its
architect, Bruce Graham of Skidmore, Owings
and Merrill, and structural engineer, Fazlur Khan,
were the same team that designed the John
Hancock Center. This structural system has been
compared to a clutch of squared cigarettes of
varying lengths bundled up, with nine indepen-
dent units strapped together. The height of the
building—1,454 feet—was set by the Federal
Aviation Administration; when constructed, it
was the world's tallest building. The steel
frame is clad with black aluminum and 16,000
bronze-tinted windows. Unlike the John Hancock Center, half of whose floors are devoted
to apartments, the Sears Tower, conceived
to house only offices, must accommodate the
bustle of rush hour. Entry and egress are
extremely limited, creating traffic jams as its
16,500 workers enter and leave the building.

BYRON C STONE · TOM WARD ·
RICHARD D GOSS · JAMES E BAILEY · RICHARD T LYNCH ·
EDDIE L SMITH ·
JAMES E HAVEMANN ·
JAMES M BECKNER ·
SCHULTZ III ·
C McEWEN Jr ·

KILBUCK · LEVERNE W MC...
CARN Jr · TERRY J NEUMEIER · PATRICK ...
NELLI · JAMES M MITCHELL Jr · TED W Q...
H EBERHARDT · ALVIN C FORNEY · GEOR...
ROBERTS · LOUIS D ROYSTON Jr · NELSO...
GILL · MERRILL L LANTRY · DALE L TOOLO...
NEVEVO ABEYTA · GEORGE W BURKHEA...
RANCH · EUGENE M JEWELL · WILLIE E ...
MES · RICHARD C MARSHALL · EDWAR...
NDSEY · RICHARD B FITZGIBBON · FRA...
B GOODWIN · GEOFFREY E GREEN · RO...
OWITZ · HARRY A HIPKE · JAMES ...
S · LARRY R TAYLOR · LEO A BAUER...
DERMAN · PAUL W MANSIR · JOSE...
JOE R MOSSMAN · JAME...
MES T KEARNS · FRANK ...
M J HENRY · KENNETH R I...
AMARENA-SALAZAR · DE...
N · HAROLD A BIRD · FRANK BOY...
...CLOTH · EDWARD H FOX · ERNE...
ERNEST L MILLER · EDGAR L PETERSO...
HELL · ROYNALD E TAYLOR · FRED R...
UESDALE · JERRY D UNDERWOOD · ...
EMBREY · EDGAR L HAWKINS · DUANE...
MANN · MANUEL FORTUNATO FERNAN...
GRAHAM Jr · GEORGE L SHOOK Jr · DAV...
ELLIS · WILLIAM E HILL · BERNARD P MU...
IALII · GARY L SCHEMEL · TROY M THOM...
D LOGAN · JESUS R MARIANO · JAMES ...
VOODS · VIRGIL GALAN CRUZ · LAUREN...
J BIENKOWSKI · ELIJAS ROUILLON...

VIETNAM VETERANS MEMORIAL

1982

WASHINGTON, D.C.

Memorials in Washington tend to invite controversy. For Maya Lin, who designed the Vietnam Veterans Memorial, the controversial nature of the war itself provoked her sunken 492-foot black granite monolith inscribed with the names of 57,939 U.S. soldiers killed or missing in the war (1959–75). Lin won the design competition for the memorial as a graduate student; the Cooper-Lecky Partnership was retained to execute the design. The monument comprises two 250-foot-long wedge-shaped walls set at a 125-degree angle, one pointing east to the Washington Monument and the other pointing west to the Lincoln Memorial. Each wall is composed of 73 black granite slabs. The names start at the top of the east wall, continue chronologically to its apex, and then begin again as the west wall emerges from the ground; they end at the beginning, where the two walls abut. Names are added at the bottom of appropriate panels as additional deaths are confirmed and status changes are made; since the memorial's dedication, on Veterans Day 1982, 252 additional names have been inscribed with a hand-held sandblasting machine, as shown here. The memorial has evoked the complexity of feelings and recollections of America's longest conflict.

189

BIG BEN RESTORATION

1985

LONDON

The clock tower of the Houses of Parliament
(1840–70, Sir Charles Barry) dates from 1860.
Its site has been England's seat of government
since King Canute (Knut the Great of Denmark),
whose reign began in 1016 A.D. Construction
of the 320-foot-tall tower and the clock itself
began in 1840, under the supervision of Sir
Benjamin Hall, a member of Parliament who,
as chief commissioner of works, had his name
etched on the side of the first tower bell. Hence,
the nickname "Big Ben" refers not to the famous
four-faced clock but to the resonant bell on
which the hours are struck. The clock, which has
kept exact time since it was set in motion in
May 1859, is the largest in Britain. Its four dials,
24 feet in diameter, and its minute hand, 14 feet
long, are fashioned of hollow copper for light-
ness. A major cleaning and restoration of the
clock and tower commenced in 1983 and
continued for 15 months with a six-person crew.
Cast-iron panels on the roof were replaced and
new gold-leaf trim applied to its ornamental
ironwork. Parts of the clock's face and hands
were also replaced during the restoration.

PYRAMIDE DU LOUVRE

1985–89

PARIS

In 1981 French President François Mitterand
requested that the eight-century-old Louvre,
located in the heart of Paris—and in the hearts
of the French people—be modernized, expanded,
and better integrated with the city, all without
compromising the integrity of the historic building.
The commission turned to I. M. Pei (1917–),
whose solution was to reorganize the U-shaped
museum around a focal courtyard with a new
main entrance for greater access and construct a
large expansion under the courtyard; the latter
allowed the exhibition space to be reorganized
and nearly doubled and serves as a bridge to
the museum's three wings. The entrance, which
leads underground, is a transparent glass
pyramid, the geometric shape that encloses the
largest area with the least volume while affording
the greatest amount of natural light. Like the
Eiffel Tower, the pyramid defies gravity, appearing
almost weightless—the antithesis of traditional
pyramids. The structure is 71 feet high at its
apex and 116 feet wide at its base. Constructed
of specially manufactured, colorless glass (the
glass alone weighs 105 tons), the pyramid floods
the lower hall with light. Its inclined walls and
transparent skin allow the broader, taller palace
beyond to be seen clearly from all angles. The
contrast of glass and stone, transparency and
opacity, allows the pyramid to complement
rather than compete with the historic Louvre.

CHANNEL TUNNEL

1987-94

FOLKESTONE, ENGLAND, TO CALAIS, FRANCE

The idea for an underwater tunnel between England and France was first proposed in 1802 by Albert Matthieu, an engineer working for Napoleon, as a strategy for invading England. Other schemes have been developed in the intervening years, but it was not until 1987 that construction began on what has become known as the Chunnel, which extends from Folkestone, England, to Calais, France. The Channel Tunnel, a 31-mile-long undersea tunnel more than 120 feet below the seabed, includes three parallel passages, two for passenger trains and a third for servicing and utilities. The tunnel's cost of more than $15 billion, twice its original estimate, will require substantial subsidies by the next generation of travelers. The trains used in the Channel Tunnel include car "ferries," special shuttle trains onto which passengers drive their automobiles and remain for the half-hour tunnel crossing; conventional passenger trains; and special trains or cars for trucks.

1989–92

BALTIMORE

At one time government buildings were typically
solid, serious places where the public's business
was conducted in suitably austere settings, often
cast in a style evocative of the Roman Forum or
the Athenian Parthenon. For these buildings
success was measured by their ability to withstand modernity and weather the ages. The state
of Maryland's most ambitious public building
project has been a $105 million, 48,000-seat
baseball-only ballpark for the Baltimore Orioles.
The 85-acre site cost an additional $99 million.
The design, by Hellmuth, Obata and Kassabaum,
contrasts sharply with ultramodern stadiums,
such as the Houston Astrodome. Camden Yards is,
in fact, a ballpark, not a stadium; its asymmetrical
design recalls historic ballparks and has started
a new wave of pseudohistoric ballpark design.
Its goal was to be compatible in scale,
configuration, and color with Baltimore's civic
buildings—hence the use of brick and steel rather
than concrete. The outer wall of the upper roof
concourse was designed so that fans can see the
city. Trusses were anchored to each other by
huge steel gusset plates. The ballpark incorpo-
rates a spectacular renovated B&O Railroad
warehouse (1905), the longest structure east of
the Mississippi River. By the standards of other
ballparks, Camden Yards is lavishly appointed,
with fountains and wrought-iron and marble
details. "A home run for baseball architecture,"
declared the *Baltimore Sun*.

U.S. HOLOCAUST MEMORIAL MUSEUM

1989-93

WASHINGTON, D.C.

When architect James Ingo Freed (1930–)
of Pei Cobb Freed and Partners joined a team
of museum professionals to create the U.S.
Holocaust Memorial Museum, the major problem
was how to tell the story of the Holocaust.
Unlike other museums, there was no collection to
house, no large legacy to fund construction, no
demonstrated demand for its creation. A group of
survivors, led by writer Elie Weisel and Baltimore
builder Harvey Meyerhoff, helped form the U.S.
Holocaust Memorial Council to organize the
project. The story of the Holocaust suggested a
narrative for the building: a prelude to the
disaster on the top floor and the continuing story
of the horrors as visitors descend to the main
floor. Freed based the memorial's design on the
architecture of the death camps themselves:
highly efficient designs reflecting the demonic
side of technology. The museum, by contrast,
is intended as a metaphor of destruction and
thus is fractured, twisted, and ominous. It is
organized internally around the three-story Hall
of Witnesses, whose skylight is warped and
eccentrically pitched; the photograph shows a
crane lifting a portion of the skylight structure.
Four five-story brick towers evoke the death
camps' sentry stations. The experience ends
in the 6,000-square-foot Hall of Remembrance,
a somber, ghettolike space for prayer and
contemplation. The building is articulated in
brick and limestone to synthesize with its
neighbors–the limestone-clad Bureau of
Engraving and Printing (1914) to the south and
the brick Auditor's Building (1879) to the north.

ENDEAVOUR

1992

HOUSTON, TEXAS

After the National Aeronautics and Space Administration landed a man on the moon in 1969, the program moved to a rocket-boosted, airplanelike vehicle that could take off and land from an airport runway; deploy, retrieve, and repair satellites in orbit; and conduct various scientific experiments, performing these tasks for up to 100 missions. Construction of the first shuttle prototypes continued through the 1970s; the *Enterprise* was carried aloft by a Boeing 747 in 1977. One persistent problem related to the tiles that served as a shield to provide heat protection during the shuttle's reentry into the atmosphere. The 33,000 tiles were too fragile and did not bond to the fuselage properly; thousands of them fell off during test flights. Each space shuttle is 120 feet long with a wing span of 80 feet, permitting as many as seven flight crew members to work and live on the flight deck. The shuttle can carry payloads of up to 65,000 pounds into orbit at an altitude of 230 miles. At 11:39 a.m. on January 28, 1986, after nine earlier routine flights, the space shuttle *Challenger* exploded at 46,000 feet, killing the seven crew members aboard, including teacher Christa McAuliffe, the first nonprofessional selected to travel into space. After this tragic accident the space program was suspended for more than two years until September 28, 1988, when the shuttle *Discovery* resumed flights. The *Endeavour*, one of the more recent space shuttles, was launched on May 7, 1992.

STATUE OF FREEDOM
RESTORATION

1993

WASHINGTON, D.C.

One hair-raising episode in the annals of restoration occurred when the newly restored seven-ton Statue of Freedom *(Freedom Triumphant in Peace and War)* was anchored to the dome of the U.S. Capitol after several months of cleaning and repairs. The 19½-foot Statue of Freedom depicts a bronze female warrior wearing an eagle helmet and clutching a sheathed sword in her right hand and a laurel wreath of victory in the other. It was designed by Thomas Crawford (1814–57) under the supervision of Thomas U. Walter and was dedicated in 1863 as an inspiration to Union troops. On December 2, 1863, the day of the statue's dedication, Confederate prisoners housed within the Capitol jeered as a crowd of several thousand, many with opera glasses, craned their necks to watch as *Freedom* was set in place. A battery of artillery from Camp Berry, stationed on the grounds east of the Capitol, fired a 35-gun salute, one volley for each state, followed by replies from the other forts around the city. The $780,000 restoration of the statue included waterblasting, chemical treatment to restore its early blue-green color and protect it from further corrosion, filling more than 700 pits with bronze plugs, and filling larger gaps with specially cast bronze plates. The statue was airlifted by an S-64F Sikorsky Skycrane, set on a specially built cone on a small platform atop the dome in a 40-mile-per-hour down draft, and bolted down.